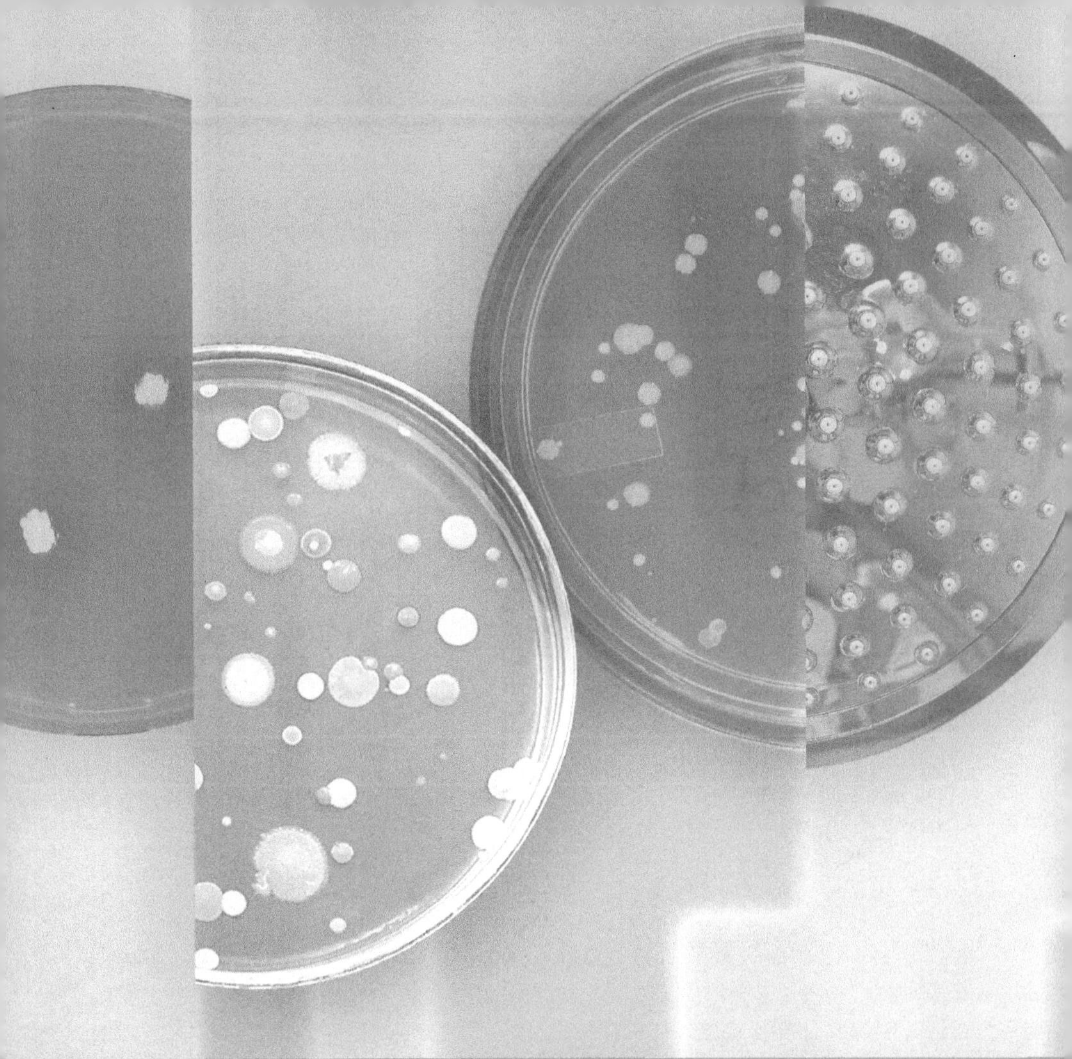

OPPORTUNISTIC PREMISE PLUMBING PATHOGENS

OPPORTUNISTIC PREMISE PLUMBING PATHOGENS

Joseph O. Falkinham, III

JENNY STANFORD
PUBLISHING

Published by

Jenny Stanford Publishing Pte. Ltd.
101 Thomson Road
#06-01, United Square
Singapore 307591

Email: editorial@jennystanford.com
Web: www.jennystanford.com

British Library Cataloguing-in-Publication Data
A catalogue record for this book is available from the British Library.

Opportunistic Premise Plumbing Pathogens

ISBN 978-981-4968-40-9 (Hardcover)
ISBN 978-1-003-32100-2 (eBook)

Contents

Preface

This book has a long history. It is an outgrowth of my 45-year study of the sources and ecology of members of the genus *Mycobacterium*; especially *Mycobacterium avium* and its related species comprising the *Mycobacterium avium* complex (MAC). As my students and I identified sources of MAC and the environmental conditions that favored their growth, survival, and persistence, we encountered other opportunistic premise plumbing pathogens (OPPPs) that seemed either to occupy the same habitats as the MAC or excluded the MAC. Those observations led to a short review, "Common features of opportunistic premise plumbing pathogens," published in the *International Journal of Environmental Research and Public Health* in 2015. Writing that review collected my thoughts and led to my carrying two lists around in my head. One was a lengthening list of opportunistic premise plumbing pathogens. The second was a list of environmental factors that were determinants of the presence or absence of OPPPs. The more I looked and read, more OPPPs and environmental determinants were added to my mental lists. Eventually I decided that there were a sufficient number of OPPPs and environmental determinants as well as a public health motivation to warrant writing a book. My purpose is to identify the common features of OPPPs and that, in turn, will lead to the development of remediation methods.

My objectives in writing this book include

- The introduction and description of known OPPPs and their public health and economic impact
- Identifying and listing the shared characteristics of known and emerging OPPPs
- Describing selective conditions of drinking water distribution systems and premise plumbing that have led to OPPPs' emergence
- Describing the microbial adaptations of OPPPs leading to survival, persistence, and growth in drinking water distribution systems and premise plumbing

The target readership is quite wide and includes

- Civil and environmental engineers in the drinking water industry
- Private and public utility operators of drinking water treatment and distribution systems
- Public health professions interested in waterborne disease
- Federal, state, and municipal regulators involved in maintaining public health and water quality
- Clinical, public health, and environmental microbiologists in universities, colleges, municipal, state, and federal public health laboratories, reference laboratories, and private consulting laboratories
- Primary care, pulmonary, and infectious disease physicians

Because of the breadth of the readership and the variety of the OPPPs, I have tried to keep scientific jargon to a minimum. Further, as I anticipate the target readers are busy, the book is short and focused. One major lesson of this book is to let the readers become aware that OPPPs surround humans and are readily transmitted to susceptible individuals. The second lesson of the book is that OPPPs are different from classic waterborne pathogens such as *Escherichia coli, Vibrio, Salmonella*, and *Shigella* that are contaminants of drinking water, not colonists like the OPPPs. That knowledge, in turn, should inform the reader that classic methods of providing safe drinking water will not work with OPPPs. I hope the information in this book will contribute to the development of novel measures to reduce exposure of people to the OPPPs.

Acknowledgments

I will be forever in debt to Professor Bruce C. Parker for it was he who led to my work with mycobacteria. In 1975, he knocked on my door and asked if I knew anything about mycobacteria. Well, fortunately, I did, having learned about mycobacteria during my four years as a clinical microbiologist in the USAF Medical Corps. Later, after the publication of our first three papers, I met established mycobacteriologists, Manny Wolinsky from Cleveland, George Kubica, and Bob Good from the Centers for Disease Control and Prevention (CDC), and Larry Wayne from the Long Beach VA Hospital at meetings. They welcomed us and encouraged our efforts and remained my mentors and advisors through the years. Early on, I met Richard C. Wallace at an American Thoracic Society meeting in Kansas City and spent a delightful afternoon talking about the environmental mycobacteria and possible projects. We continue talking and collaborating to this day.

It was Richard LeChevallier of the American Water Company who introduced me to drinking water and drinking water distribution systems. We met at an EPA goal-setting meeting at U.C. Irvine and were fortunate to receive funding to look for mycobacteria in drinking water systems. As we now know, drinking water systems and household plumbing are colonized by mycobacteria. The introduction to drinking water led me to send an email to an author of a paper in the *Journal of the American Water Works Association* describing microorganisms in premise (household) plumbing. As I typed the email address, I got to @vt and discovered the author Marc Edwards was at my University, Virginia Tech. We got together over chicken fingers and a marvelous collaboration has flowered. There are days that I feel I was standing at a train station and my arm was caught by Amtrak's Acela and I went from zero to 100 miles per hour in a second. The collaboration with Marc and his colleagues Amy Pruden and Andrea Dietrich, and Bill Knocke has been wonderful. Their questions and projects have enriched my knowledge of drinking water and of the opportunistic premise plumbing pathogens (OPPPs) that colonize our drinking water.

Led by another mentor, Mike Iseman at National Jewish Hospital in Denver, I was able to show that isolates of *Mycobacterium avium* in the showerhead of an *M. avium*-infected patient were identical with the patient's isolates. When I gave a talk about that work to Norm Pace's laboratory gang at the University of Colorado, Norm sent members of his lab across the United States to sample showerheads, not by culture, but by rDNA gene sequence analysis. That work demonstrated that members of the genus *Mycobacterium* were the most frequent (70%) of showerhead samples and led to my concept that humans are surrounded by mycobacteria. Further, that work led to the discovery that another common waterborne bacterium, namely the salmon-pink colored colonies of *Methylobacterium*, was present when the mycobacteria were absent. That discovery by Norm's group led me to start considering other waterborne opportunistic pathogens.

Chapter 1

Introduction to OPPPs

Opportunistic premise plumbing pathogens (OPPPs) are waterborne microorganisms that are normal inhabitants of premise plumbing and drinking water and cause infections in individuals with predisposing conditions. The best-known examples of OPPPs include bacteria: *Legionella pneumophila, Pseudomonas aeruginosa, Mycobacterium avium* complex (MAC), and other nontuberculous mycobacteria (NTM), *Stenotrophomonas maltophilia, Acinetobacter baumannii, Sphingomonas paucimobilis, Methylobacterium* spp., *Segniliparus* spp., *Cupriavidus* spp., yeast *Candida* spp, and protozoa *Acanthamoeba* spp. and *Vermamoeba* spp. (Table 1.1). They are normal inhabitants of natural soils and waters (e.g., lakes, rivers, streams, and ponds) and drinking water distribution systems. These microorganisms share a number of features that allow them to survive, persist, and grow in premise plumbing; the pipes in houses, hospitals, apartments, condominiums, and office buildings. Premise plumbing includes the plumbing (i.e., water-delivery systems) in houses, apartments, condominiums, hospitals, offices, factories, and other workplaces.

As humans are surrounded by OPPPs, infections can be traced to a community or hospital exposures. Here, the focus is on drinking water, not sewage-contaminated water, as OPPP infections have been traced, almost exclusively, to drinking water.

Opportunistic Premise Plumbing Pathogens
Joseph O. Falkinham, III

ISBN 978-981-4968-40-9 (Hardcover), 978-1-003-32100-2 (eBook)
www.jennystanford.com

OPPPs grow under the low carbon conditions (i.e., oligotrophic) and occasional stagnation (i.e., low oxygen) of drinking water. One reason for OPPP-persistence and survival in premise plumbing is its ability to adhere to surfaces where they grow to form biofilms, which are cells growing in an extracellular matrix of polysaccharides, proteins, DNA, and lipid. The attached extracellular matrix and OPPP cells allow them to avoid being washed out by the flow of water and further, survive exposure to disinfectants (e.g., chlorine) due to the impenetrability of disinfectants through the layers of matrix material and cells.

Table 1.1 Opportunistic premise plumbing pathogens (OPPPs)

Existing OPPPs
Legionella pneumophila and other *Legionella* spp.
Pseudomonas aeruginosa
Burkholderia cepacia
MAC and NTM
Stenotrophomonas maltophila
Acinetobacter baumannii
Sphingomonas paucimobilis
Segniliparus spp.
Cupriavidus spp.
Candida spp.
Acanthamoeba (Vermamoeba) spp.

Infections caused by the OPPPs include pneumonias involving the lungs or appearing in the blood or skin. OPPP infections have been linked directly to water or aerosol exposures or indirectly to medical solutions (e.g., for skin sterilization) or devices (e.g., bronchoscopes or heater-coolers). OPPPs are opportunistic pathogens, meaning they infect and cause disease in individuals with preexisting conditions (i.e., risk factors). The risk factors could be a consequence of behavior (e.g., smoking or occupation), a pre-existing condition (e.g., cystic fibrosis), or simply residence in a hospital or long-term care facility. Patients in intensive care units, including children, and patients with long-term use of indwelling catheters are especially at risk. Risk factors for OPPP infection include prior lung damage, cancer, chemotherapy, immunodeficiency, and inherited conditions such as cystic fibrosis. One other group, taller, slender, and older men and women are particularly susceptible to infection caused by *M. avium* and other NTM. To date, there has been no identification of the physiologic basis for the increased susceptibility of the taller, slender, older individuals.

For clarification, the OPPPs will always be referred to as colonists of premise plumbing, not contaminants. The OPPPs are introduced into premise plumbing from piped water supplies, one of their natural habitats and are thereby colonists; simply occupying a new habitat. Contaminants originate from other sources (e.g., human or animal feces). The OPPPs reside and grow in natural waters and soils. Humans have provided them with a new habitat, piped water systems for which they are perfectly adapted. Thus, all humans have done is provide the OPPPs with a new habitat to occupy. For example, OPPPs occupy and grow in the water before it is introduced into treatment plants, distribution systems, and premise plumbing. I reserve "contaminant" for those microorganisms that can be introduced into water supplies, yet fail to grow and disappear over time. Those, such as *Escherichia coli*, are found in high numbers at the source of the contamination event, and numbers fall as the distance from the source increases. That is the basis for point source identification, which follows increasing upstream to identify the source. In contrast to contaminating microorganisms like *E. coli*, the OPPPs increase in number from the source. That is the reverse of what we see with classical waterborne pathogens, such as *E. coli*. For *E. coli* and other classical waterborne pathogens, the water is simply a carrier (vector), as the microbes do not grow. Further, OPPP numbers and their frequencies of isolation do not correlate with indices of fecal contamination of drinking water (Falkinham et al., 2001). Drinking water that is described as free from *E. coli* or other fecal coliforms, may not be free of OPPPs, so caution must be exercised in evaluating water quality solely by *E. coli* or fecal coliform numbers. Rather, OPPPs grow in water, so water serves as both a vector for transmission and distribution, and a site for amplification through growth. For example, NTM numbers increase two-fold upon their transmission in a drinking water distribution system (Falkinham et al., 2001).

Here, I focus on a limited number of waterborne opportunistic pathogens as exemplary OPPPs in order to identify the traits and characteristics that they exhibit and share in common. The shared characteristics identified are those that lead to OPPP colonization of premise plumbing. One aim of this work is to identify those common features to provide guidance to the development of measures to reduce their numbers in premise plumbing. The hypothesis is that a number of measures can be taken, for example raising water heater temperatures that would reduce exposure to any of the OPPPs. It has been pointed out that any remediation strategy must consider the range of OPPPs and should not focus on a single water-borne opportunistic pathogen, such as *L. pneumophila* (Kanamori et al., 2016). Thus, by identifying the common features of OPPPs, strategies can be developed to reduce different OPPP species simultaneously. In that fashion, remediation strategies do not need to be distinct and separate for each OPPP.

Little data is available accurately reporting the incidence and prevalence of OPPP infections. The best data concern is *Legionella pneumophila,* as these infections must be reported. Recently published studies show that the incidence of cases of *Legionella* spp. is less than 10,000 per year (CDC, 2018). Estimates show there are 32,600 cases of *Pseudomonas aeruginosa* infection annually (CDC, 2019) and 16,000 NTM cases annually over the period 1998–2005 (Billinger et al., 2009). However, the data for *P. aeruginosa* and NTM are estimating infections caused by those and other OPPPs, saving *L. pneumophila,* and are not required to be reported. Some progress is being made for NTM infections with 11 states now requiring reporting of NTM infections. A further problem with assessing the impact of OPPP infection is that many cases of disease caused by the OPPPs may not require hospitalization and may thereby go unreported.

The annual economic cost of those infections is quite high in the United States, over $430 million for *Legionella* spp. and $425 million for NTM infections (Collier et al., 2012). In another study involving analysis of national medical claims maintained by the

U.S. Center for Medicare and Medicaid Services for infections caused by *L. pneumophila, Pseudomonas* spp., and NTM identified 617,291 cases over the period 1999–2006 that resulted in an annual cost of $600 million (Naumova et al., 2016; Mirsaeidi et al., 2015). In Canada, France, Germany, and the United Kingdom, the average direct medical costs per person-year for *M. avium* complex lung disease (only) were: $16,200 (Canada), €11,600 (Germany), €17,900 (France), and £9,700 (United Kingdom) (Goring et al., 2018). The estimates for costs are likely underestimated due to a number of reasons. First, only *L. pneumophila, Pseudomonas* spp., and *Mycobacterium* spp. were included, not other OPPPs like *S. maltophilia* and *A. baumannii.* Second, the costs are likely underestimated as the infections reported were only those involving hospitalization and do not include infections that were treated outside of hospitals. Finally, those cases may not include the cost of litigation involved in nosocomial infections caused by any of those three. For example, there have been recent reports of *Mycobacterium chimaera* infections associated with cardiac surgery (Sax et al., 2015). The infections have been tracked to *M. chimaera* (a relative of *M. avium*) in a heater-cooler instrument used to warm or cool the patient and blood during surgery. The infections have been reported from throughout the world and have been traced to the introduction of *M. chimaera* from the water used to test the operation of the instruments before leaving the factory (Haller et al., 2016). As the number of operations employing the heater-cooler throughout the world is quite high, and the infections carry with them high mortality, many suits have been filed in the United States. In like fashion, cervical lymph node infections in children caused by NTM have been traced the use of NTM-colonized water in dental units (Hatzenbuehler et al., 2017; Singh et al., 2018).

An integrated analysis of the two simultaneous trends of increasing OPPP prevalence and the increasing numbers of more susceptible individuals are needed in order to identify strategies for better protecting public health now and in the future. It is the objective of this review to: (a) identify and describe the characteristics of premise plumbing (Chapter 3), (b) describe the characteristics of OPPPs (Chapters 4 and 5), (c) identify the common characteristics of OPPPs that lead to their persistence and growth (Chapter 6), (d) introduce newly emerging OPPPs (Chapter 7), (e) identify and describe the conditions in premise plumbing that lead to OPPP selection in drinking water systems and premise plumbing (Chapter 8), (f) describe methods for OPPP monitoring (Chapter 9), (g) describe the sources and consequences of OPPP presence (Chapter 10), and (h) discuss the challenge of OPPPs to current water treatment paradigms (Chapter 10). It is hoped that this integrated and comprehensive review of OPPPs will inform methods for identifying and providing the rationale for measures to reduce numbers of OPPPs and exposure to OPPPs in premise plumbing (Chapter 11). Toward those ends, the review will employ current knowledge of representative OPPPs as exemplars and not as independent, unique waterborne pathogens.

References

Billinger ME, Olivier KN, Viboud C, Montes de Oca R, Steiner C, Holland SM, Prevots DR. 2009. Nontuberculous mycobacteria-associated lung disease in hospitalized persons, United States, 1998–2005. *Emerg Infect Dis.* 15: 1562–1569.

Centers for Disease Control and Prevention (CDC). 2018. Legionnaires' Disease Surveillance Summary Report, United States, 2014–2015. https://www.cdc.gov/drugresistance/pdf/threats-report/pseudomonas-aeruginosa-508.pdf

Centers for Disease Control and Prevention (CDC). 2019. Multidrug-resistant *Pseudomonas aeruginosa*. https://www.cdc.gov/drugresistance/pdf/threats-report/pseudomonas-aeruginosa-508.pdf

Collier SA, Stockman LJ, Hicks LA, Garrison LE, Zhou FJ, Beach MJ. 2012. Direct healthcare costs of selected diseases primarily or partially transmitted by water. *Epidemiol. Infect.* 140: 2003–2013.

Falkinham JO III, Norton CD, LeChevallier MW. 2001. Factors influencing numbers of *Mycobacterium avium, Mycobacterium intracellulare,* and other mycobacteria in drinking water distribution systems. *Appl. Environ. Microbiol.* 67: 1225–1231.

Goring SM, Wilson JB, Risebrough NR, Gallagher J, Carroll S, Heap KJ, Obradovic M, Loebinger MR, Diel R. 2018. The cost of *Mycobacterium avium* complex lung disease in Canada, France, Germany, and the

United Kingdom: A nationally representative observational study. *BMC Health Services Res.* 18: 700. doi: 10.1186/s12913-018-3489-8.

Haller S, Höller C, Jacobshagen A, Hamouda O, Abu Sin M, Monnet DL, Plachouras D, Eckmanns. 2016. Contamination during production of heater-cooler units by *Mycobacterium chimaera* potential cause for invasive cardiovascular infections: Results from and outbreak investigation in Germany, April 2015 to February 2016. *Euro Surveill.* 21: 30215.

Hatzenbuehler LA, Tobin-D'Angelo M, Drenzek C, Peralta G, Cranmer LC, Anderson EJ, Milla SS, Abramowicz S, Yi J, Hilinski J, Rajan R, Whitley MK, Gower V, Berkowitz, Shaprio CA, Williams JK, Harmon P, Shane AL. 2017. Pediatric dental clinic-associated outbreak of *Mycobacterium abscessus* infection. *J Ped Infect Dis Soc.* 6: e116–e122.

Kanamori H, Weber DJ, Rutala WA. 2016. Healthcare outbreaks associated with water reservoir and infection prevention strategies. *Clin Infect Dis.* 62: 1423–1435.

Mirsaeidi M, Allen MB, Ebrahimi G, Schraufnagel D. 2015. National hospital costs for pulmonary mycobacterial disease in the US from 2001 to 2012. *Int J Mycobateriol.* 4: 156–157.

Naumova EN, Liss A, Jagai JS, Behlau I, Griffiths JK. 2016. Hospitalizations due to selected infections caused by opportunistic premise plumbing pathogens (OPPP) and reported drug resistance in the United States older adult populations in 1991–2006. *J Publ Hlth Policy.* 37: 500–513.

Sax H, Bloemberg G, Hasse B, Sommerstein R, Kohler P, Achermann Y, Rössle M,. Falk V, Kuster SP, Böttger EC, Weber R. 2015. Prolonged outbreak of *Mycobacterium chimaera* infection after open-chest heart surgery. *Clin Infect Dis.* 61: 67–74.

Singh J, O'Donnell K, Ashouri N, Adler-Shohet FC, Nieves D, Tran MT, Arrieta A, Tran L, Cheung M, Zahn M. 2018. Outbreak of invasive nontuberculous *Mycobacterium* (NTM) infections associated with a pediatric dental practice. *OFID.* 5: S29.

Chapter 2

Public Health Impact of OPPPs

Evidence is emerging that both the number of OPPPs in drinking water and the number of individuals who are at risk of OPPP infection are increasing. Thus, we need to expand our knowledge beyond the limited number of model pathogens reviewed here to others; for example, *Stentrophomonas maltophila* and *Acinetobacter baumannii*. As these will be discussed later (Chapter 10), the normal inhabitants of drinking water systems and premise plumbing challenge the current paradigm for water treatment. Specifically, the OPPPs colonize, grow, and persist in spite of current water treatment methods.

Over the period 2000–2009, the incidence rate of *Legionella* spp. infections increased almost 200%, from 0.39 cases per 100,000 persons in 2000 to 1.15 in 2009 (CDC, 2017). For *P. aeruginosa*, although yearly incidence figures are not available because reporting is not required, a total of almost 11,000 hospital-acquired bloodstream, pneumonia, and urinary tract infections were identified in the United States over the period of January 1992 through May 1999 (CDC, 2019). For NTM, both the estimated number of isolates and cases have increased. Specifically, the frequency of isolation of NTM from clinical specimens has risen from 9.1 to 14.1/100,000 over the period 1997–2003; a mean annual increase of 8.4% (Billinger et al., 2014). Pulmonary

Opportunistic Premise Plumbing Pathogens
Joseph O. Falkinham, III

ISBN 978-981-4968-40-9 (Hardcover), 978-1-003-32100-2 (eBook)
www.jennystanford.com

NTM hospitalizations increased from 2.1 to 2.4 per 100,000 in Florida; a 3.2% annual increase (Billinger et al., 2014). The increase NTM pulmonary disease is predominantly in women (6.5% per year in Florida and 4.6% in New York (Billinger et al., 2014). The majority of NTM infections in the United States are caused by one species, *Mycobacterium avium*.

The OPPPs also have a significant impact on the mortality of diseases that can be transmitted by water (Table 2.1). From a total of 6,939 annual total deaths in the United States over the period of 2003–2009, 6,301 (91%) were associated with infections caused by *Pseudomonas*, NTM, and *Legionella*, all OPPPs (Gargano et al., 2017). Only 7% of deaths were linked to pathogens (e.g., *Salmonella*, Hepatitis A) transmitted via the fecal-oral route (Gargano et al., 2017). Average annual deaths numbered 250 for *Legionella*, 1,216 for NTM, and 4,835 for *Pseudomonas* (Gargano et al., 2017). That data strongly suggests that increased attention should be placed on the OPPPs and their infections.

Table 2.1 Estimated frequencies and costs of OPPP infections

	Estimated infection (reference)	
OPPP	**Frequency**	**Cost ($)**
L. pneumophila	50,000 (1)	500 million (2)
P. aeruginosa	11,000 (3)	(2)
M. avium	180,000 (4)	500 million (2)

(1) CDC (2011).

(2) Collier et al. (2012).

(3) Arduino et al. (2014).

(4) Billinger et al. (2014).

Most believe that the numbers of cases for each of the three OPPPs are underestimated. First, infections may not be reported. Second, community-acquired infections may not be counted as the individuals may not be hospitalized. Third, infections caused by *P. aeruginosa* and NTM, unlike *Legionella* spp. infections, are not required to be reported. Thus, for *P. aeruginosa* and NTM infections, the best available data on prevalence and incidence of cases are estimates. For *P. aeruginosa* and NTM infections, hospital records have been surveyed using disease codes to identify cases. However, such an estimate is woefully inadequate, especially for NTM infections, as a majority of patients are not hospitalized and are seen, diagnosed, and treated as outpatients. If we include an estimate of the proportion of NTM infected patients who have not been hospitalized (e.g., 5 to 1), then there are approximately 150,000-NTM-infected individuals in the United States.

Increases in both the proportion of older individuals and those that are immunodeficient or carry other risk factors for OPPP infection are major contributors to the increases in OPPP infection. For example, the proportion of United States individuals over 60 years is estimated to reach 25% by 2025. The effect of an aging population on OPPP infection and disease is illustrated by the six-fold increased likelihood of *M. avium* lung disease in the United States from 15/100,000 for the general population versus 100/100,000 amongst individuals over 60 years (Billinger et al., 2014). The same trend holds true for *L. pneumophila* infection (CDC, 2019). In like fashion, as the number of immunodeficient individuals continue to rise (e.g., cancer, chemotherapy, and transplantation patients), who are more susceptible to OPPP infection, so the incidence of OPPP infection will increase. Changes in building plumbing practices and disinfection emphasizing water conservation might also be increasing the likelihood of OPPP occurrence and disease.

It is possible to identify and outline some of the problems posed by the emergence of OPPPs drinking water systems and premise plumbing (Table 2.1). They include: (a) the loss of confidence in the safety and microbial quality of utility-delivered water, (b) the threat of increased regulatory burdens on drinking water utilities, (c) the possibility of mandates (i.e., regulations and installation of equipment) to hospitals and long-term care facilities to protect patients and residents from OPPPs, and (d) the possibility of additional costs to owners of home and condominiums to install and maintain equipment to reduce exposure to OPPPs.

Let me provide one example of the complexity of issues that might face homeowners and operators of hospitals and long-term care facilities. OPPPs in drinking water have led to the installation of remediation measures, such as point-of-use (POU) devices for improvement of water quality. The impact of POU devices (i.e., microfiltered/granular activated carbon-containing and reverse-osmosis water dispensers) on drinking water bacterial numbers was measured at Italian residential healthcare facilities for the elderly (Sacchetti et al., 2015). The study revealed that, as expected, chlorine concentrations were reduced by both types of POU devices, and conductivity, calcium, sodium, hardness, and total dissolved solids were reduced by the reverse-osmosis device (Sacchetti et al., 2015). Sadly, rather than reducing bacterial counts, the use of point-of-use filters led to increased bacterial numbers in the treated water compared to the pre-POU water (Sacchetti et al., 2015). Further, two opportunistic pathogens, *P. aeruginosa* and *S. maltophila*, were detected in the water from the POU devices (Sacchetti et al., 2015). Such evidence suggests caution in the installation of POU devices as their ability to remove disinfectants, ions, or solids is not an indicator of their ability to reduce bacterial numbers. In fact, the devices may put susceptible individuals at risk for OPPP infection.

Table 2.2 Anticipated public health challenges of emerging OPPPs

Loss of Confidence in the Quality of Utility-Delivered Water
Burden of Additional Regulations Governing Water Treatment
Burden of Additional Treatment of Drinking Water
Increasing Stringency of Guidelines for OPPPs in Drinking Water
Increasing Cost of Water Treatment by: Utilities, Hospitals, Apartments and Condominiums, and Homes
Lack of Standards for OPPP Numbers in Drinking Water

Recognition of OPPPs and their potential to cause chronic or life-threatening infections should drive two regulatory issues. First, diseases caused by OPPPs must become reportable; otherwise, public health officials will be unable to craft and then execute strategies to reduce OPPP presence and exposure to susceptible individuals. Second, standards for "acceptable" water must be written, reviewed, and made public. For example, in the United

States, the acceptable level of fecal coliforms in drinking water is 1 per 100 mL of a water sample. What should the standard be for the different OPPPs? In addition, should the standard for OPPP numbers in water and plumbing of hospitals be stricter than the standard for homes and other residences? Such standards should be based on consideration of the dose-response of each OPPP. Unfortunately, the dose-response for *M. avium*, the prominent NTM in the United States water is simply unknown. Further, it might prove to be the case that certain genotypes of *M. avium* are of more concern for human health than others. Without that data, municipal, state, and federal agencies will be unable to provide guidance. Thus, a research priority is to define those dose-response relationships for each OPPP. In part, this has started as the *M. avium* complex (MAC) is listed in the fourth edition of the EPA "Candidate Contaminant List" (EPA, 2015). Inclusion on that list indicates that EPA requires further research for those; in fact, it is a list of priority contaminants that requires research.

There is one major benefit to the recognition of OPPPs, in spite of the fact that they represent different taxonomic groups of microorganisms. Even though they are quite different in structural and genetic makeup, they share common features that may lead to methods of reducing the numbers of all OPPPs in drinking water.

References

Billinger ME, Olivier KN, Viboud C, Monets de Oca R, Steiner C. Holland SM, Prevots DR. 2009. Nontuberculous mycobacteria-associated lung disease in hospitalized persons, United States, 1998–2005. *Emerg Infect Dis.* 15: 1562–1569.

Collier SA, Stockman LJ, Hicks LA, Garrison LE, Zhou FJ, Beach MJ. 2012. Direct healthcare costs of selected diseases primarily or partially transmitted by water. *Epidemiol Infect.* 140: 2003–2013.

Centers for Disease Control and Prevention. 2017. National Notifiable Diseases Surveillance System, Annual Tables of Infectious Disease Data. Atlanta, GA. CDC Division of Health Informatics and Surveillance, 2016. https://www.cdc.gov/nndss/infectious-tables.html.

CDC. 2019. Multidrug-resistant *Pseudomonas aeruginosa.* https://www.cdc.gov/drugresistance/pdf/threats-report/pseudomonas-aeruginosa-508.pdf.

Environmental Protection Agency. 2015. EPA Candidate Contaminant List. EPA, Cincinnati, OH; EPA-HQ-OW-2012-0217 at regulations.gov.

Gargano JW, Adams EA, Collier SA, Fullerton KE, Feinman SJ, Beach MJ. 2017. Mortality from selected diseases that can be transmitted by water–United States, 2003–2009. *J. Water Health.* 15: 438–450.

Sacchetti R. De Luca G, Guberti E, Zanetti F. 2015. Quality of drinking water treated at point of use in residential healthcare facilities for the elderly. *Int J Environ Res Public Health.* 12: 11163–11177; doi: 10.3390/ijerph120911163.

Yoder JS, Eddy BA, Wisvesvara GS, Capewell L, Beach MJ. 2010. The epidemiology of primary amoebic meningoencephalitis in the USA, 1962–2008. *Epidemiol Infect.* 138: 968–975.

Chapter 3

Characteristics of Premise Plumbing

3.1 Introduction

Premise plumbing, the plumbing in homes, apartments, condominiums, hospitals, offices, and manufacturing buildings has a number of unique characteristics that are determinants of the persistence and proliferation of waterborne opportunistic plumbing pathogens, the opportunistic premise plumbing pathogens (OPPPs). These characteristics are listed and their ramifications for the OPPPs in Table 3.1. As the sources of infection by OPPPs include drinking water and premise plumbing, a consideration of the characteristics of premise plumbing is necessary.

Depending on the characteristics of premise plumbing, resident numbers of OPPPs can be low or quite high. For example, the highest numbers of *Mycobacterium* spp. that have been recovered by culture [i.e., colony-forming units (CFU) per milliliter of water or per square centimeter of biofilm] and have been recovered from buildings with recirculating hot water systems, such as apartments and condominiums (Tichenor et al., 2012). Although there have been a few systematic studies of OPPP numbers and their frequency of isolation of OPPPs in hospitals, it is likely that those recirculating systems harbor high OPPP numbers. As most patients infected with OPPPs have a transient (e.g., surgical

Opportunistic Premise Plumbing Pathogens
Joseph O. Falkinham, III

ISBN 978-981-4968-40-9 (Hardcover), 978-1-003-32100-2 (eBook)
www.jennystanford.com

trauma, antibiotic-therapy), a long term (e.g., COPD, cancer, or cancer), or a genetic predisposing condition (e.g., cystic fibrosis, immunodeficiency), it follows that they are subject to repeated infections, even after antibiotic therapy successfully results in the disappearance of disease symptoms (Wallace et al., 1998). As the source of OPPPs can be premise plumbing, it follows that the numbers of OPPPs in the system, called the "colonization pressure", which is a determinant of infection. For example, high numbers of *M. avium* or *L. pneumophila* in premise plumbing would lead to high numbers of both in shower aerosols (Parker et al., 1983) and thus individuals run the risk of shower-associated infection. What follows is a factor-by-factor description of each premise plumbing feature and the ramifications of each on OPPP numbers (Table 3.1).

Table 3.1 Characteristics and ramifications of premise plumbing

Characteristic	Ramification(s)
Large Surface and Volume	Large Numbers of OPPPs
High Surface-to-Volume Ratio	Large Area for Biofilms
	Reduce Disinfectant Concentration
Water Heater	Amplifier of OPPP Numbers
	Warm Water in System for Growth
Different Pipe Materials	Favorable for Adherence/Biofilm
	Inhibit Some OPPPs
	Stimulate Growth of Some OPPPs
Plumbing Aged	Corrosion Stimulates OPPP Growth
	Corrosion Releases OPPP Toxins
Water Hardness	Cations for OPPP Adherence
	Cations/Anions Inhibit OPPP Growth
	Cations/Anions Stimulate Growth
Water Nutrients	AOC Stimulates OPPP Growth
	Nutrients Generate OPPP Diversity
High Water Residence Time Increase in OPPP Numbers	Increase in OPPP Metabolites
Variable Flow Rate	Flow Rate Influences Adherence
	High Flow Rate = Biofilm Release

Low Disinfectant Residual	Release OPPPs from Inhibition
Use of Plastic Pipes	Leaching of Organics as AOC
	Increased Growth of OPPPs
	Increased OPPP Biofilms
Dead Ends	Refuges for OPPPs
	Inoculation of Plumbing
Resident Microbiome	Stimulation of Microbes
	Inhibition of Microbes

3.2 Premise Plumbing Surface Area

The first characteristic of premise plumbing is to consider its length and surface area. An average house in the United States has 50 ft. of ¾" (1.9 cm) diameter and 50 ft. of ½" (1.3 cm) diameter pipe. As most measurements of microbial numbers are expressed as cells per square centimeter of biofilm, the length measurements are transformed to the centimeter, equaling 3,048 cm length of household plumbing. The total inside surface area equals 9,000 cm^2 for ¾" (1.9 cm) diameter pipe and 6,000 cm^2 for ½" (1.3 cm) diameter pipe; a total of 15,000 cm^2 (2,325 in^2) for typical house. Using that value for total pipe surface area (independent whether it is copper, stainless steel, plastic, etc.), if measurements show 5,000 cells of a particular OPPP per square centimeter of pipe surface area of a representative biofilm sample, a typical house would have a total of 75 million OPPP cells.

A like calculation of the total volume of a typical home leads to an estimate of 6,340 cm^3 (1.67 gal). If the number of OPPP cells per cube centimeter was equal to 1,000/cm^3, the premise plumbing water would contain 6.3 million cells, smaller than one-tenth of the cells in the biofilm.

These numbers are simply estimated and subject to wide variation as each house will have a different mix of pipe types, each with its unique ability to support a biofilm of different OPPPs (Mullis and Falkinham, 2013). Although time-consuming, it is possible to calculate a unique set of values based on the characteristics of the premise plumbing in any building. That calculation would include consideration of different composition percentages (e.g., 50% copper and 50% polyvinyl chloride (PVC) or 50% copper and 50% stainless steel) and the number of OPPP cells in biofilms on these different surfaces. A consumer could consider selecting plumbing pipes that will support the lowest number of adherent OPPPs.

In conclusion, households have high numbers of OPPPs, most of which are in biofilms on the inside surface of the pipes. Second, as the volume of the hot water heater dominates the volume of the premise plumbing (see Section 3.4 below), it is likely that the OPPP composition and numbers of the hot water heater will be the determinant of that in premise plumbing.

3.3 High Surface-to-Volume Ratio

The diameter of premise plumbing pipes [e.g., ½" (1.3 cm) and ¾" (1.9 cm) in diameter) is considerably smaller than that of distribution system pipes [e.g., 8" (20.3 cm)]. Therefore, the ratio of surface area to volume of an 8" (20.3 cm) drinking water distribution pipe is 0.20, for a ¾"(1.9 cm) pipe is 2.6, and for a ½" (1.3 cm) pipe is 3.2. That means the surface-to-volume ratio for premise plumbing pipes is between 13- and 16-fold greater than distribution system pipes. It follows that the much larger surface-to-volume ratio of premise plumbing pipe offers a greater area for microbial adherence and biofilm formation. Further, in a flowing system, as is premise plumbing, most microbes will be washed out unless they adhere. Even assuming that the same biofilm OPPP cell densities are the same for the distribution system and premise plumbing per square centimeter of pipe surface, there will be many more OPPPs in premise plumbing because of the higher surface-to-volume ratio. The greater shear forces generated in the smaller diameter premise plumbing is unlikely to reduce biofilm formation, as it has been shown that shear does increase biofilm formation (Lehtola et al., 2007; Torvinen et al., 2007). As *L. pneumophila* (Cooper and Hanlon, 2000), *P. aeruginosa* (Drenkard, 2003), and *M. avium* (Steed and Falkinham, 2003), cells in biofilms are more resistant to killing by disinfectants, even the residual disinfectant concentrations in premise plumbing will be unable to reduce OPPP numbers.

There are other ramifications of the high surface-to-volume ratio in premise plumbing. First, a high surface-to-volume ratio means that there is more pipe surface for interaction with disinfectants. As pipe materials can accelerate chlorine and chloramine decay (e.g., copper), disinfectants disappear relatively rapidly in premise plumbing with their higher surface-to-volume ratio (Zheng, 2013; Zheng et al., 2015). The higher surface area means the heat will be dissipated faster, so water heated to equal or above the temperature leading to cellular death (e.g., ≥ 55 °C) will cool, allowing microbial survival. In addition, a greater surface area means that leaching from pipe materials, particularly metals or nutrients from plastic pipes (Connell et al., 2016; Neu and Hammes, 2020), will be higher in premise plumbing leading to an increase in the numbers of biofilm microorganisms (Proctor et al., 2016; Neu and Hammes, 2020).

3.4 Water Heater

In addition to the volume of water in the pipes, it is necessary to include the volume of the household water heater, a glass-lined thermos of some 30 gallons (114,000 cm^3). Based simply on volume, the water heater is the dominant (i.e., 18-fold greater volume) water reservoir of premise plumbing. Using a value of 1,000 OPPP cells/cm^3, the water heater can contain 114 million OPPP cells. Clearly, the composition and number of microorganisms in premise plumbing are heavily influenced by the number and types in the hot water heater.

A water heater is a glass-lined, insulated thermos, and the water is heated by a gas flame or electric coil. The nature of the heating element directly influences microbial numbers and growth. For example, stratification occurs in electric-heated water heaters and the highest numbers of *Legionella pneumophila* were found are particular locations corresponding to their optimal temperature for growth (Alary and Joly, 1991).

Depending on the water heater's set temperature, microorganisms will be killed or their growth stimulated. The microbial flora in a water heater is strongly influenced by temperature (Dai et al., 2018). Generally, the water heater is a microbial culture system, increasing the numbers of OPPPs in premise plumbing (Hilborn et al., 2006). Growth is a consequence of the fact that biological reactions are increased two-fold by every 10 °C increase in temperature. Measurements of *Mycobacterium* spp. numbers in households of *Mycobacterium* spp.-infected patients showed that *Mycobacterium* spp. isolation frequency and numbers were significantly higher in household water samples collected from homes in which the water temperature was ≤125 °F (50 °C). In contrast, homes with a hot water heater temperature of ≥130 °F (55 °C) seldom had *Mycobacterium* spp. (Falkinham, 2011).

3.5 Premise Pipe Materials

Compared to a distribution system, premise plumbing is far more variable in composition. Copper, brass, stainless steel, galvanized steel, and plastic [e.g., PVC, high-density polyethylene (HDPE), and cross-linked polyethylene (PEX)], and glass (i.e., hot water heater) can all be in premise plumbing. The effects of pipe material composition can be direct or indirect. Direct effects have to do with surface physical characteristics (e.g., topology and hydrophobicity), while indirect effects are due to the release of antimicrobial or growth-stimulating compounds.

Adherence to premise plumbing materials varies widely, even for a single microbial species or strain. For example, 10-fold more *M. avium* cells adhere to galvanized compared to copper pipe material (Mullis and Falkinham, 2013). An indirect effect is observed in copper pipe, as chlorine concentrations fall rapidly in a copper pipe (Zheng et al., 2015), leading to the release of chlorine inhibition of microorganisms. Another indirect effect is responsible for the low numbers of *P. aeruginosa* on copper pipes. Cells of *P. aeruginosa* are sensitive to released copper and enter a viable, but unculturable (VBNC) state that leads to low recovery of *P. aeruginosa* cells in biofilms (Bédard et al., 2016).

Plastic pipes are being increasingly used in premise plumbing, substituting for copper, in particular. The plastics include HDPE, polypropylene (PP), PVC, and cross-linked polyethylene (PEXa, PEXb, and PEXc). With the exception of PP, organic matter as assimilable organic carbon (AOC) can support the growth of microorganisms migrating from the plastic pipes to the water (Connell et al., 2016; Neu and Hammes, 2020). That migrated organic matter stimulates bacterial growth and population diversity (Proctor et al., 2016). Although growth stimulation has been observed, there have been no measurements of the growth of specific bacteria belonging to the group of OPPPs.

3.6 Premise Plumbing Age

The age of premise plumbing is a factor influencing microbial numbers. Long service times of premise plumbing are associated with higher levels of microorganisms compared to the plumbing of short service life (Zheng, 2013). This could be due to the long service plumbing as having a larger biofilm, a greater likelihood of releasing growth-stimulatory metals or organics, or having more sites available for microbial colonization or a combination.

3.7 Water Hardness and Nutrient

Water hardness is the concentration of minerals, largely Ca^{2+} and Mg^{2+}, and nutrient in drinking water varies across the United States. Both might influence OPPPs in premise plumbing. Primarily, the effect of the nutrient is on the growth of OPPPs, whereas hardness would affect both growth (essential micronutrients) and the adherence of OPPPs to premise plumbing pipes. As cations in laboratory studies have been shown to affect adherence and biofilm formation (Carter et al., 2003), water hardness ought to influence OPPP numbers in premise plumbing. Unfortunately, the research performed has not distinguished between (a) the effect of cations on growth and (b) the effect of cations on adherence of OPPPs. The cations Ca^{2+}, Mg^{2+}, or Zn^{2+}, but not Fe^{2+}, increased the number of *M. avium* cells in biofilms two-fold after 14 days (Carter et al., 2003). As the number of adherent *M. avium* cells was approximately the same after 1 day and 7 days incubation (Carter et al., 2003), it would appear that the effect of the cations was on the growth of adherent cells, rather than their attachment to the surface. The effect of the cations was independent of concentration over the range 0.1 μM–10 μM (Carter et al., 2003), but biofilm formation was not measured for combinations of cations. Thus, it is not known whether the cations might have had distinct roles or acted additively or synergistically.

Measurements of biofilm formation of *Mycobacterium fortuitum* and *Mycobacterium chelonae* provided the surprising discovery that biofilm area and height were the same under high (laboratory medium) and low (sterile tap water) nutrient conditions on HDPE surfaces (Hall-Stoodley et al., 1999). The addition of phosphorous (10 μg/L) to a recirculating biofilm-reactor increased the number of heterotrophic bacteria in biofilms, but decreased *M. avium* numbers (Torvinen et al., 2007). The authors speculated that *M. avium* loss was due to competition, namely heterotrophs responded more quickly to the phosphorus addition and occupied available sites for adherence (Torvinen et al., 2007).

3.8 Water Usage and Variable Flow Rates

Water flow rates in premise plumbing can vary markedly. In a home, flow rates are high at two periods of the day; morning and early evening. At other times, there may be no flow at all. Morning and evening are also the times when hot water may be circulating. When there is no flow and the absence of mixing, oxygen is consumed by the resident OPPPs and heterotrophic microorganisms (as many as 10^5 cells/mL) leading to stagnation. Thus, in a home, apartment, or condominium, there is a strong selection for microorganisms able to grow at low oxygen levels. Studies have shown that low water usage rates are associated with high microbial numbers compared with premise plumbing subject to high water usage rates (Zheng, 2013; Zheng et al., 2015). This is particularly a problem for buildings that have been relatively unoccupied during the COVID-19 pandemic. At the opposite end of the spectrum, at high usage rates, there is more opportunity for microorganisms to be released from biofilms into the flowing water. Such release may not occur at low usage rates and reflects the resistance of biofilms to dissipation as a result of high shear forces under rapid flow, turbulent conditions (Lehtola et al., 2007).

Adherence of OPPPs to premise plumbing surfaces does not require the absence of flow. High, turbulent water flow rates (e.g., 183 mL/h, Reynold's Number = 15,000) did not prevent the formation and persistence of cells of *L. pneumophila* (15 days) and *M. avium* (28 days) in biofilms in a recirculating reactor (Lehtola et al., 2007). In a similar study of water flow velocity and biofilm formation counts of *M. avium* cells in biofilms were 2–3 times higher at a water flow rate of 0.24 m/s compared to 0.1 m/s (Torvinen et al., 2007). Although the available data establish that water flow velocity influences adherence to surfaces and biofilm formation, whether there are optimal flow rates for OPPPs has not been established.

3.9 Water Residence Time

Residence time is the length of time water is retained in a pipe before it flows out of a tap. Compared to a distribution system, the residence time in premise plumbing is quite long. This means that there is more time for OPPPs to adhere to surfaces where they form biofilms, more time for OPPPs to grow in water, and more time for disinfectants to interact with the pipe surfaces. As the interaction of disinfectants with pipe surfaces leads to disinfectant disappearance, the length of time OPPPs and other microorganisms are exposed to disinfectants in premise plumbing is relatively short.

Recirculation could also be thought of as relevant to residence time by having an effect due to exposure to water in premise plumbing. Although rare in single-family homes, recirculating systems are common in buildings with apartments, condominiums, or offices and in hospitals. The highest number of nontuberculous mycobacteria my lab has encountered (i.e., ≥ 100,000 CFU/mL) has been from samples collected from apartment and condominium buildings with recirculating hot water systems (Tichenor et al., 2012). In a comparative study of hot water systems, it was shown that recirculating systems had a greater volume of water at temperatures conducive for OPPP growth if the temperature was set at 49 °C (Brazeau and Edwards, 2013). Tanks in recirculating systems also had lower concentrations of disinfectant residual and 3–20-times more sediment (Brazeau and Edwards, 2013). Interestingly, if the water temperature was set at 60 °C, there was less opportunity for growth of OPPPs, as the temperature was kept high enough to restrict OPPP growth for a longer period of time (Brazeau and Edwards, 2013).

3.10 Low Disinfectant Residual

Premise plumbing characteristically has a low disinfectant concentration, referred to as the disinfectant residual. A low disinfectant residual may not be sufficient to kill a proportion of microorganisms in the water or on surfaces (biofilms). As the disinfectant concentration falls during water transit in the distribution system, it is not uncommon to measure increases—called regrowth—in microbial numbers, especially for OPPPs like *Mycobacterium* spp. (Falkinham et al., 2001). Thus, by the time the water enters the home, hospital, elder care center, or residential and commercial buildings, the concentration of disinfectant is reduced to the point of lacking any ability to reduce the numbers or prevent the growth of the disinfectant-resistant OPPPs in premise plumbing. Independently, disinfectant concentration falls as a result of corrosion. Depending on the age of premise plumbing, corrosion of pipes can be high; expressed as the Larson Iron Corrosion Index (Larson and Skold, 1958).

3.11 Organics and Metals Leach from Premise Plumbing

The organic matter in premise plumbing is that portion of the total organic matter not degraded during movement from the treatment plant and through the distribution system. The microbial-degradable fraction of the organic matter is called AOC. The AOC measurement is not a chemical measurement, but it is a functional measure based on the growth response of standard strains of bacteria. For plastic pipes, the AOC consists of plasticizers and unreacted agents for polymerization. The major component of the organic matter entering premise plumbing is degradation-resistant organic matter. A major portion of that consists of humic and fulvic acids, the dark complex compounds resulting from partial degradation of the dark structural matter in plants. Interestingly, the growth of one OPPP, *M. avium,* is stimulated by humic and fulvic acids (Kirschner et al., 2005). A survey of the ability of other known and emerging OPPPs is warranted to determine whether a general characteristic of these opportunistic pathogens is their ability to grow on humic and fulvic acids.

Metals, such as the divalent cations Fe^{2+}, Cu^{2+}, Mo^{+2}, Ni^{2+}, and Zn^{2+}, are required for microbial growth. Growth promotion occurs at low concentrations, as these metals are inhibitory at high concentrations. Premise plumbing can be a source for these metals; particularly under conditions of corrosion. Thus, corrosion, in addition to reducing disinfectant concentration, can increase the concentration of divalent cations in premise plumbing. If metal concentrations are limiting, corrosion will lead to increased growth of OPPPs in premise plumbing. In Flint, Michigan, the absence of corrosion-control led to the increase in the number of *Legionella pneumophila* in the distribution system. However, it is important to point out that the concentration of Cu^{2+}, Ni^{2+}, and Zn^{2+} is important as high concentrations can lead to growth inhibition. The heavy metals Hg^{2+}, Cd^{2+}, and Pb^{2+} are uniformly anti-microbial as, unlike the other cations, they have no essential function in microbial cells. Inhibition of microbial growth is a consequence of their substitution for other divalent cations in proteins and enzymes leading to loss of activity. Corrosion can also result in the release of the toxic heavy metals. Rather than growth-stimulation as a result of increased concentration of Fe^{2+}, Cu^{2+}, Ni^{2+}, and Zn^{2+} in water, the release of the toxic heavy metals will inhibit the growth of microorganisms in premise plumbing. Interestingly, two established OPPPs, *Mycobacterium* spp. (Falkinham et al., 1984) and *Pseudomonas aeruginosa* (Teitzel and Parsek, 2003) are relatively tolerant of heavy metals in either suspension or biofilms. *M. avium* is particularly resistant; growing at heavy metal concentrations 10-fold higher than other bacteria (Falkinham et al., 1984). Perhaps, a common character of OPPPs is the tolerance of heavy metals.

3.12 Stagnation

As the flow in premise plumbing can be zero for long periods of time, especially in residential systems in homes, apartments, and condominiums, resident OPPPs and other microorganisms will respire, reduce oxygen concentrations, even to the point of creating microaerobic (6% oxygen) or anaerobic (0% oxygen) conditions. Such reduced oxygen concentrations may not prevent the continued metabolism and growth of OPPPs. *P. aeruginosa* is able to grow anaerobically using nitrate as a terminal electron-acceptor (Fewson and Nicholas, 1961). *M. avium* is capable of growth at 12% and 6% oxygen. In fact, *M. avium* growth at 12% oxygen was equal to that in the air (21% oxygen, Lewis and Falkinham, 2015), suggesting that for *M. avium*, stagnation is not a limit to growth and persistence in premise plumbing.

3.13 Dead Ends

Dead ends represent areas of no flow and thus can become refuges for OPPPs that can re-infect premise plumbing after measures have been taken to disinfect the system (e.g., hyper-chlorination or heat-shock).

3.14 Convective Mixing in Distal Pipes

A controlled, pilot-scale of a hot water system showed that pipe configurations, namely slanted to promote convective mixing and vertical to limit convective mixing demonstrated that convective mixing promoted the growth of *Legionella pneumophila* (Rhoads et al., 2016). It is likely that the convective mixing promoted the distribution of nutrients throughout the pipe volume and prevented temperature stratification, leading to uniform temperatures that were close to the optimal temperature for growth of *L. pneumophila* (Rhoads et al., 2016).

3.15 Resident Microbiome in Premise Plumbing

A recent study has shown that surface adherence and biofilm formation are strongly influenced by the existing surface biofilm (Khweek and Amer, 2018). Biofilm formation of *L. pneumophila* was the highest on surfaces with an existing "normal" flora, a surface population formed by exposure to a drinking water sample (Khweek and Amer, 2018). Interestingly, biofilms that included *Methylobacterium* spp. were more likely to have high numbers of *L. pneumophila*, compared to biofilms comprising relatively a few microorganisms (Rogers et al., 1994). By contrast, Feazel et al. (2009), using a non-cultural approach to measuring microbial numbers in showerheads, disclosed the interesting interaction between *Methylobacterium* spp. and *Mycobacterium* spp. In that study, it was shown that the presence of *Methylobacterium* spp. in a biofilm was associated with the absence of *Mycobacterium* spp. and the reverse. The non-culture approach chosen by Feazel et al. (2009) is quite strong, as it provides a broad survey of biofilms in showerheads across the United States and was not biased by an inability to cultivate a particular genus of bacteria. We followed up the report of the mutual inhibition of biofilm formation, measuring the numbers of *Mycobacterium* spp. and *Methylobacterium* spp. biofilm samples of premise plumbing in homes of patients in Montgomery County, Pennsylvania with *M. avium* pulmonary disease. The results documented the mutual exclusion of the two OPPPs (Falkinham et al., 2018). Measurements of *M. avium* adherence to *Methylobacterium* spp. biofilms, and vice-versa, showed that either could inhibit the adherence of the other to biofilms (Munoz-Agea et al., 2019). Evidently, the first to colonize a surface can exclude the other.

References

Alary, M., Joly, J.R. 1991. Risk factors for contamination of domestic hot water systems by Legionellae. *Appl. Environ. Microbiol.* 57: 2360–2367.

Bédard, E., Prévost, M., Déziel, E. 2016. *Pseudomonas aeruginosa* in premise plumbing of large buildings. *Microbiol. Open.* 5: 937–956.

Brazeau, R.H., Edwards, M.A. 2013. Role of hot water system design on factors influential to pathogen regrowth: Temperature, chlorine residual, hydrogen evolution, and sediment. *Environ. Engnr. Sci.* 30: 617–627.

Carter, G., Wu, M., Drummond, D.C., Bermudez, L.E. 2003. Characterization of biofilm formation by clinical isolates of *Mycobacterium avium. J. Med. Microbiol.* 52: 747–752.

Connell, M., Stenson, A., Weinrich, L., LeChevallier, M., Boyd, S.L., Ghosal, R.R., Dey, R., Welton, A.J. 2016. PEX and PP water pipes: Assimilable carbon, chemicals, and odors. *J Am Water Works Assoc.* 108: E192–E204.

Cooper, I.R., Hanlon, G.W. 2010. Resistance of *Legionella pneumophila* serotype 1 biofilms to chlorine-based disinfection. *J. Hosp. Infect.* 74: 152–159.

Dai, D., Rhoads, W.J., Edwards, M.A., Pruden, A. 2018. Shotgun metagenomics reveals taxonomic and functional shifts in hot water microbiome due to temperature setting and stagnation. *Front Microbiol.* 9: 2695. doi: 10.3389.fmicb.2018.02695.

Drenkard, E. 2003. Antimicrobial resistance of *Pseudomonas aeruginosa* biofilms. *MIcrob. Infect.* 5: 1213–1219.

Falkinham III, J.O. 2011. Nontuberculous mycobacteria from household plumbing of patients with nontuberculous mycobacteria disease. *Emerg. Infect. Dis.* 17: 419–424.

Falkinham III, J.O., George, K.L., Parker, B.C., Gruft, h. 1984. In vitro susceptibility of human and environmental isolates of *Mycobacterium avium, M. intracellulare,* and *M. scrofulaceum* to heavy metal salts and oxyanions. *Antimicrob. Agents Chemother.* 25: 137–139.

Falkinham III, J.O., Williams, M.D., Kwait, R., Lande, L. 2016. *Methylobacterium* spp. as an indicator for the presence or absence of *Mycobacterium* spp. *Intl. J. Mycobacteriol.* 5: 240–243.

Fewson, C.A., Nichols, J.D. 1961. Nitrate reductase from *Pseudomonas aeruginosa*. *Biochim. Biophys. Acta*. 49: 335–349.

Hall-Stoodley, L., Keevil, C.W., Lappin-Scott, H.M. 1999. *Mycobacterium fortuitum* and *Mycobacterium chelonae* biofilm formation under high and low nutrient conditions. *J. Appl. Microbiol. Symp. Suppl.* 85: 60S–69S.

Khweek, A.A., Amer, A.O. 2018. Factors mediating environmental biofilm formation by *Legionella pneumophila*. *Front. Cell. Infect. Microbiol.* 8: 38. https://doi.org/10.3389/fcimb.2018.00038.

Larson, T.E., Skold, R.V. 1958. Laboratory studies relating mineral quality of water to corrosion of steel and cast iron pipe. *Corrosion*. 14: 43–46.

Lehtola, M.J., Torvinen, E., Kusnetsov, J., Pitkänen, T., Maunula, L., von Bonsdorff, C.-H., Martikainen, P.J., Wilks, S.A., Keevil, C.W., Miettinen, I.T. 2007. Survival of *Mycobacterium avium, Legionella pneumophila, Escherichia coli,* and calciviruses in drinking water-associated biofilms grown under high-shear turbulent flow. *Appl. Environ. Microbiol.* 73: 2854–2859.

Lewis, A.H., Falkinham III, J.O. 2015. Microaerobic growth and anaerobic survival of *Mycobacterium avium, Mycobacterium intracellulare* and *Mycobacterium scrofulaceum*. *Int. J. Mycobacteriol.* 4: 25–30.

Mullis, S.N, Falkinham III, J.O. 2013. Adherence and biofilm formation of *Mycobacterium avium, Mycobacterium intracellulare* and *Mycobacterium abscessus* to household plumbing materials. *J. Appl. Microbiol.* 115: 908–914.

Muńoz-Egea, M.C., Ji, P., Pruden, A., Falkinham III, J.O. 2017. Inhibition of adherence of *Mycobacterium avium* to plumbing surface biofilms of *Methylobacterium* spp. *Pathogens*. 6: 42.

Neu, L., Hammes, F. 2020. Feeding the building plumbing microbiome: The importance of synthetic polymeric materials for biofilm formation and management. *Water*. 12: 1774.

Proctor, C.R., Gächter, M., Kötzsch, S., Rölli, F., Sigrist, R., Wlaser, J.-C., Hammes, F. 2016. Biofilms in shower hoses: Choice of pipe material influences bacterial growth and communities. *Env Sci Water Res Technol*. 2: 670–682.

Rhoads, W.J., Pruden, A., Edwards, M.A. 2016. Convective mixing in distal pipes exacerbates *Legionella pneumophila* growth in hot water plumbing. *Pathogens*. 5: 29 doi:10.3390/pathogens5010029.

Rogers, J., Dowset, A.B., Dennis, P.J., Lee, J.V., Keevil, C.W. 1994. Influence of temperature and plumbing material selection on biofilm formation and growth of *Legionella pneumophila* in a model potable water system containing complex microbial flora. *Appl. Environ. Microbiol.* 60: 1585–1592.

Steed, K.A., Falkinham III, J.O. 2006. Effect of growth in biofilms on chlorine susceptibility of *Mycobacterium avium* and *Mycobacterium intracellulare*. *Appl. Environ. Microbiol.* 72: 4007–4100.

Tichenor, W.S., Thurlow, J., McNulty, S., Brown-Elliott, B.A., Wallace Jr., R.J., Falkinham III, J.O. 2012. Nontuberculous mycobacteria in household plumbing as possible cause of chronic rhinosinusitis. *Emerg. Infect. Dis.* 18: 1612–1617.

Teitzel, G.M., Parsek, M.R. 2003. Heavy metal resistance of biofilm and planktonic *Pseudomonas aeruginosa*. *Appl. Environ. Microbiol.* 69: 2313–2320.

Torvinen, E., Lehtola, M.J., Martikainen, P.J., Miettinen, I.T. 2007. Survival of *Mycobacterium avium* in drinking water biofilms as affected by water flow velocity, availability of phosphorous, and temperature. *Appl. Environ. Microbiol.* 73: 6201–6207.

Zheng, M. 2013. Factors contributing to chlorine decay and microbial presence in drinking water following stagnation in premise plumbing. Master's Thesis, University of Tennessee. http://trace.tennessee.edu/utk_gradthes/2485.

Zheng, M, He, C., He, Q. 2015. Fate of free chlorine in drinking water during distribution in premise plumbing. *Ecotoxic.* 24: 2151–2155.

Chapter 4

Established OPPPs

This chapter is a brief review of the shared characteristics of three recognized opportunistic premise plumbing pathogens (OPPPs), namely *Legionella pneumophila, Pseudomonas aeruginosa* (and *Burkholderia* spp.), and *Mycobacterium avium* complex, and amoeba and protozoa. The following sections (i.e., Sections 4.1 through 4.4) are descriptions of the infections, physiology, and sources of the recognized OPPPs. These sections are not meant to be exhaustive; for example, little is presented on their mechanisms of virulence. Following this chapter, there are two: one describing OPPPs as amoebae-resisting microorganisms (Chapter 5, OPPPs as ARMs) and one describing the common features of OPPPs (Chapter 6). In total, those chapters are meant to provide information to allow one to determine whether a novel microbial isolate from a patient could possibly be OPPPs. Those lessons are employed to identify and describe the emerging OPPPs in another chapter (Chapter 7). Further, my objective is to identify common features of OPPPs to identify sources and develop methods for remediation and thereby reduce nosocomial or community infection. Doubtless, some readers may feel particular microorganisms should have been included, but my intent has not been to be comprehensive, but to identify common features of OPPPs to guide the inclusion of other microbial pathogens and to lead to the identification of sources and methods for remediation.

Opportunistic Premise Plumbing Pathogens
Joseph O. Falkinham, III

ISBN 978-981-4968-40-9 (Hardcover), 978-1-003-32100-2 (eBook)
www.jennystanford.com

Although these microbial species belong to different taxonomic groups, they share a number of characteristics in common that lead to a shared ecology. Common features of established OPPPs are: (a) the source of infection has been traced to drinking water from premise plumbing, (b) growth or regrowth in a drinking water distribution system or premise plumbing, (c) growth or survival under stagnant or anaerobic conditions, (d) disinfection-resistance, (e) surface adherence and biofilm formation, and (f) survival and growth in amoebae (Table 4.1). As can be appreciated, those features ensure that OPPPs are well adapted to persistence, growth, and survival in drinking water distribution systems and premise plumbing.

Table 4.1 Common features of established opportunistic premise plumbing pathogens

Drinking Water Source
Growth or Regrowth in Drinking Water
Disinfectant-Resistance
Surface Adherence
Biofilm Formation
Desiccation-Tolerance
Survival and Growth in Amoebae
Numbers do Not Correlate with *E. coli* or Fecal Coliforms

4.1 *Legionella pneumophila*

4.1.1 *L. pneumophila* Epidemiology and Diseases

L. pneumophila is likely the most widely known waterborne opportunistic premise plumbing pathogen (OPPP). It causes life-threatening pneumonia, Legionnaires' disease, whose sources have been traced to natural waters, drinking waters, and aerosols from hot tubs, showers, and cooling towers. The estimated incidence of Legionnaires' disease is increasing at a rate of 5% per year and there were approximately 50,000–70,000 cases, of which 80% were community-acquired (CDC, 2017). Aerosols from showers (Hayes-Phillips et al., 2019) and cooling towers (Fitzhenry et al., 2017) are not the only sources as Legionnaires' disease was linked to the presence of *L. pneumophila* in a dental unit waterline (Ricci et al., 2012).

4.1.2 Risk Factors for *L. pneumophila* Disease

Independent risk factors for Legionnaires' disease include exposure to a non-municipal (private) water supply, recent premise plumbing repairs, and smoking (Straus et al., 1996). Having an electric water heater, working more than 40 h per week, and spending nights away from home before the onset of illness were additional risk factors (Straus et al., 1996). The risk of Legionnaires' disease was significantly higher for New York City workers in transportation, repair, protective services, cleaning, or construction compared to people in other occupations (Farnham et al., 2014). Also, as outbreaks of Legionnaires' disease have been linked to its presence in cooling towers (Fitzhenry et al., 2017), close proximity to cooling towers are a likely risk factor. Weather, specifically warm [60–80 °F (16–27 °C)] and humid (> 80% relative humidity) were associated with higher Legionnaires' disease probability, suggesting a basis for the lower incidence and prevalence of Legionnaires' disease in the arid southwestern United States (Simmering et al., 2017).

4.1.3 *L. pneumophila* Sources and Ecology

4.1.3.1 Sources of *L. pneumophila*

Since the first reports of Legionnaires' disease, cooling towers have been offered as the source (Fitzhenry et al., 2017). However, cooling towers are not the only sources and searches for outbreak sources have often been misdirected. I first learned about other environmental sources of Legionnaires' disease when an outbreak was traced to a hot tub (spa) display at a do-it-yourself store near Virginia Tech (Hershey et al., 1997). In the Netherlands, *L. pneumophila* in a whirlpool spa at a flower show had the same genotype as those of 28 or 29 culture-positive patients (Den Boer et al., 2002). *L. pneumophila* has been cultured from water collected from retirement homes (37% of samples) and group homes (10% of samples) in Rome, Italy (De Filippis et al., 2018). *L. pneumophila* has been detected in cold water taps throughout the United States (Donohue et al., 2014) and in two water distribution systems in Australia (Whiley et al., 2014). *L. pneumophila* infection is not just exposure to cooling towers.

In a study of hospital-acquired *L. pneumophila* infection in Flint, Michigan in 2014–2015, the outbreak was associated with recovery of *L. pneumophila* from some patients and hospital water whose isolates shared the same whole-genome multilocus sequence typing profiles (Smith et al., 2019). The hospital water system was shown to be at least one source, as measures taken to reduce *L. pneumophila* numbers coincided with the reduction in the number of cases (Smith et al., 2019). Additional sources included Flint River water and close proximity to Flint cooling towers as shown by statistical analysis (Smith et al., 2019).

L. pneumophila is present in dental unit water lines, though the percentage recovery by culture varies widely: 0.4% of 266 samples from dental offices in London and rural Northern Ireland (Pankhurst et al., 2003), 33% of 99 samples from dental clinics at the University of the Witwatersrand, South Africa (Singh and Coogan, 2005), and 36% of 50 samples from dental clinics at the University of Dentistry, Palacky University and Hospital, Olomouc, the Czech Republic (Juraskova et al., 2017). The recovery values by culture are likely underestimated as *L. pneumophila* detection by qPCR yields far higher estimates; specifically, of 86 samples, 100% were positive by qPCR versus 7% by culture (Ditommaso et al., 2016).

4.1.4 Ecology of *L. pneumophila* in the Built Environment

4.1.4.1 Premise plumbing temperature

The presence of *Legionella* in household plumbing is strongly influenced by both the type of water heater and water temperature (Alary and Joly, 1991). Specifically, in Quebec, homes with an oil or gas water heater did not yield *Legionella* (0/33); only homes with an electric water heater (69/178, 39%) yielded *Legionella* (Alary and Joly, 1991). The basis for more frequent detection of *Legionella* in electric water heaters may be due to both stratification and the lower temperatures of hot water at the bottom (30.3 °C) compared to that of fossil fuel water heaters (49.2 °C). A recent study showed that the presence of *Legionella* in hot water return lines compared to distal lines had a sensitivity of 55% and specificity of 96%, suggesting that temperature comparisons could be used as a surrogate for *Legionella* colonization (Pierre et al., 2019).

The percentage of hot water samples from instantaneous hot water devices were lower (6.2%) than from standard water heaters (30%), although the data were likely influenced by the fact that water temperatures at taps were higher in the instantaneous devices (>60 °C), than standard water heaters (≤ 50 °C) (Martinelli et al., 2000).

Legionella spp. isolation from single-family homes (8%) was significantly lower than from apartments with a centrally heated water supply (25%), suggesting that a central, recirculating hot water system, as in hospitals, condominiums, and apartments, might support greater persistence of *Legionella* (Marrie et al., 1994).

4.1.4.2 Premise plumbing chlorine

Household colonization of premise plumbing by *L. pneumophila* was associated with lower chlorine concentrations (0.35 mg/L versus 0.56 mg/L), as well as lower water heater temperatures [42 °C (107 °F) versus 47 °C (117 °F)] in premise plumbing (Straus et al., 1996).

L. pneumophila has been isolated or detected (using PCR-based approaches) in samples of drinking water collected across the United States (Donohue et al., 2014) and Canada (Marrie et al., 1994). Recovery of *Legionella* spp. from drinking water is not uniform, even within a relatively small geographic area. For example, samples of drinking water in 6 of 7 Halifax, Nova Scotia hospitals yielded *Legionella* spp., but samples from only 3 of 32 non-Halifax hospitals did not (Marrie et al., 1994).

Persistence of *Legionella* spp. in premise plumbing is aided by its ability to adhere to pipe surfaces and form biofilms.

L. pneumophila was able to colonize and form biofilms on PVC, polypropylene, steel, and stainless steel (Rogers et al., 1994a). *L. pneumophila*, as well as other waterborne microorganisms, forms biofilms on copper in low numbers (Rogers et al., 1994b; Lu et al., 2014). Interestingly, *L. pneumophila* cells survived exposure to 50 °C on plastic surfaces (Rogers et al., 1994b).

4.1.4.3 Oxygen levels in premise plumbing

Oxygen levels are likely regulators of presence or absence and numbers of *L. pneumophila*. *L. pneumophila* will grow at dissolved oxygen concentrations of between 6.0 mg/L and 6.7 mg/L; however, not at levels of 2.2 mg dissolved oxygen per liter (Wadowsky et al., 1985). Sadly, the effects on *L. pneumophila* growth and survival at intermediate concentrations of oxygen were not reported. It is known that *L. pneumophila* proliferates in "stagnant" water (Ciesielski et al., 1984), but the oxygen concentrations have not been measured in such conditions.

The presence of aerators also influences *L. pneumophila* colonization of premise plumbing. In a Denver hospital, *L. pneumophila* was isolated from 22 of 30 faucet aerators but not from 26 nonobstructed taps (Cieseilski et al., 1984).

4.1.5 Long-Term Survival of *Legionella*

Although *L. pneumophila* is fastidious and difficult to cultivate, high numbers are found in drinking water samples (Donohue et al., 2014). *L. pneumophila*'s fastidious requirements for the growth seem inconsistent with its presence in drinking water, but can be explained by its long term survival associated with morphological changes (Paszko-Kolva et al., 1992) and its ability to grow in phagocytic amoebae; falling in the category of amoebae-resisting microorganisms (ARM) (Bouyer et al., 2007). The persistence of *L. pneumophila* in drinking water may be due to its growth in amoebae; quite possibly, its presence in drinking water may require its carriage in amoebae. That hypothesis has yet to be directly tested, but if proven that will have a major impact on efforts to control this waterborne pathogen. Specifically, control of *L. pneumophila* could be exerted through the reduction in amoebae numbers. Consistent with its presence in drinking water, *L. pneumophila* is relatively chlorine-resistant (Kuchta et al., 1985). Further, water-adapted cells of *L. pneumophila* are 7-fold more resistant to chlorine than cells grown in a laboratory medium (Kuchta et al., 1985). *L. pneumophila* is found in biofilms on a variety of pipe surfaces (Rogers et al., 1994a; Rogers et al., 1994b; Lu et al., 2014). Coupled with its presence in amoebae and biofilms, the habitats occupied by *L. pneumophila* provide it with protection from disinfection, nutrients for growth, and resistance to being washed out by water flow.

The hardiness of *L. pneumophila* was demonstrated by its survival for over two years in the wastewater aeration pond of a wood-based chemical factory in Norway (Olsen et al., 2010).

4.1.6 Strain Typing *L. pneumophila*

4.1.6.1 Methods for strain typing

A variety of methods have been tested and employed for typing isolates of *L. pneumophila* during outbreaks and source tracking. Three methods for typing, namely large fragment typing (PFGE-RFLP), amplified fragment length polymorphism (AFLP), and arbitrarily primed PCR (AP-RCR or RAPD), were compared for reproducibility and discrimination using a collection of *L. pneumophila* isolates from two German hospitals (Jonas et al., 2000). All three methods yielded identical results for environmental and patient isolates that had been linked epidemiologically. AP-PCR was the least discriminatory method and suffered from a lack of reproducibility. AFLP yielded the highest interassay reproducibility (90%) and concordance in comparison with PFGE-RFLP (Jonas et al., 2000). Another typing method was disclosed in an article identifying sources of *L. pneumophila* infection in Flint, Michigan, 2014–2015 based on whole-genome multilocus sequence typing (Smith et al., 2019). Typing by variable number tandem repeat (VNTR-RFLP) analysis demonstrated high typability, reproducibility, stability, and epidemiological concordance (Pourcel et al., 2007). As VNTR-RFLP yields data that can be compiled in an internationally available database, it would appear to be the method of choice for typing *L. pneumophila*, along with whole-genome sequencing. VNTR-RFLP typing analysis led to the identification of different types in a drinking water system, their localization in different portions, and speculation about whether the presence of a specific type led to high *Legionella* counts in cold water (Rodríguez-Martínez et al., 2015). In the absence of whole-genome analysis, the extent of sequence differences between VNTR-RFLP types could be small or large. That knowledge would influence any discussion of special or evolutionary differences. To illustrate that fact, whole-genome sequencing of *L. pneumophila* isolates from persistent hotel colonization led to the discovery of 16 recombination events responsible for 98% of the single nucleotide polymorphisms (SNPs) identified in the core genome (Sânchez-Busó et al., 2014). These recombination events had the effect of increasing the genetic distance between isolates and thereby accelerating the rate of evolution of the *L. pneumophila* (Sânchez-Busó et al., 2014).

4.1.6.2 Genome evolution during *L. pneumophila* infection

Whole-genome analysis of *L. pneumophila* outbreak isolates over time has provided evidence of evolution during infection. They also call to light the possible weaknesses of other typing methods and raise the issue of the identity of different *L. pneumophila* isolates. First, it should be understood that following introduction into a new habitat, for *L. pneumophila* the human lung or premise plumbing, mutations will occur that would change DNA sequences and possibly fingerprints. Thus, it is important to know the range of whole-genome sequence differences for each typing method as has been accomplished for *Mycobacterium avium* (Lande et al., 2018).

In a study of *L. pneumophila* isolates recovered throughout persistent colonization of an Italian hotel using whole-genome sequencing, a single type was shown responsible for a majority of infections (Sânchez-Busó et al., 2016). Isolates of that *L. pneumophila* type were recovered from patients, a spa pool, and from other sites within the hospital. In addition to the discovery of the major sequence type, other related sequence types were also discovered and their lineage and relatedness to the main sequence type shown graphically (Sânchez-Busó et al., 2016). The study provided evidence of significant intra-sequence type variability that would have normally confounded any determination of the relatedness of the patient and environmental isolates.

4.1.7 Transmission of *L. pneumophila*

Legionella spp. and *L. pneumophila* are transmitted via aerosols. Transmission can occur via aerosols generated by showerheads and hot-water faucets (Bollin et al., 1985) and humidifiers (Woo et al., 1986). Hot water faucets were likely chosen as hot water favors the presence of *L. pneumophila.* The association of *L. pneumophila* presence in a hospital cooling tower with an outbreak of Legionnaires' disease provided a useful method for the calculation of exposures (Brown et al., 1999). An outbreak of Legionnaires' disease associated with an air scrubber at a wood-based chemical factory in Norway, led to the cultivation of *L. pneumophila* from the factory's aeration ponds for waste (Olsen et al., 2010). Evidently, the air-scrubber transferred aerosols of *L. pneumophila.*

There is one report of person-to-person transmission of Legionnaires' disease involving a 48-year old man who contacted *L. pneumophila* from a cooling tower and his 74-year old mother who cared for him during the initial 5-days of his infection (Coreia et al., 2016). The mother had not been to the industrial site and she and her son shared small non-ventilated rooms lacking air-conditioning (Coreia et al., 2016). Six days following the appearance of Legionnaires' disease symptoms in the son, the mother was hospitalized for symptoms of Legionnaires' disease (Coreia et al., 2016). *L. pneumophila* isolates from both patients were shown to have identical whole-genome sequences (Coreia et al., 2016). Water collected from the shared residence did not yield *Legionella* spp.

4.1.8 Controlling *L. pneumophila* in Premise Plumbing

A recent review, "Managing *L. pneumophila* in Water Systems" (LeChevallier, 2020) has systematically addressed and evaluated the various measures that can be taken to keep *L. pneumophila* in check. In addition to reminding the reader that a plan for dealing with the discovery of *L. pneumophila* in a water distribution system or building be developed before any outbreaks occur, a valuable list of guidelines from the Allegheny County Health Department, PA for interpreting *L. pneumophila* numbers and levels demanding actions is included (LeChevallier, 2020).

In a 2016 CDC study, a majority of Legionnaires' disease outbreaks (85%; 23/27) were ascribed to maintenance deficiencies, most due to insufficient disinfectant concentration (Garrison et al., 2016). Growth conditions and biofilm surface both influence susceptibility to disinfection (Saby et al., 2005; Buse et al., 2019). *L. pneumophila* is susceptible to chlorine with 99% of cells killed by exposure to 0.25 mg chlorine per liter (0.25 ppm) in 10 min (Kuchta et al., 1985). However, the species exhibit the characteristic of reduced susceptibility by water-acclimated cells (99% killing in 60–90 min) (Kuchta et al., 1985). The increased chlorine-resistance of water acclimated cells is also the case for *M. avium* (Taylor et al., 2000). In both cases, the switch appears to be an adaptation, not mutation, as water-acclimated cells lose their chlorine resistance upon cultivation in laboratory agar medium (Kuchta et al., 1985). Substitution of monochloramine for chlorine appears to reduce *L. pneumophila* prevalence in premise plumbing samples of 96 buildings from 19.8% (19/96) to 6% (6.2%) (Moore et al., 2006). Unfortunately, the substitution led to the increased recovery of *Mycobacterium* spp. from the same buildings (Casini et al., 2014; Springthorpe et al., 2015).

As stagnant water in pipes can lead to the proliferation of *L. pneumophila* and the installation of electronically timed flow taps in combination with chlorination have been tested in hospitals to determine that combination's effect on *L. pneumophila* (Totaro et al., 2020). The rationale for the installation of the timed taps is that they release water from a hospital distribution system to ensure constant disinfectant concentration. Their placement in regions of low flow or dead ends leads to increased flow, reduction of water age, and restoration of disinfectant concentration. In a hospital in Italy, the installation of timed water taps in conjunction with chlorine disinfection in four hospitals in Italy led to the disappearance of Culturable *L. pneumophila* between 1 day and 15 days (Totaro et al., 2020).

In addition to chlorine and monochloramine, other disinfectants and combinations have been tested to measure their efficacy in reducing *L. pneumophila* presence in premise plumbing. These include intermittent copper-silver ionization (Liu et al., 1998; Cloutman-Green et al., 2018), a combination of hot water, ultraviolet light, silver ions, and chlorine (Miyamoto et al., 2000), and a combination of hydrogen peroxide with silver salts (Girolamini et al., 2019). All are not totally effective for eradication, perhaps because of the increased disinfectant resistance of *L. pneumophila* cells in biofilms (Lu et al., 2014).

References

Alary M, Joly JR. 1991. Risk factors for contamination of domestic hot water systems by *Legionellae*. *Appl Environ Microbiol.* 57: 2360–2367.

Arnow PM, Weil D, Para MF. 1985. Prevalence and significance of *Legionella pneumophila* contamination of residential hot-tap water systems. *J Infect Dis.* 152: 145–151.

Baron JL, Harris JK, Holinger EP, Duda S, Stevens MJ, Robertson CD, Ross KA, Pace NR, Stout JE. 2015. Effect of monochlorine treatment of the microbial ecology of *Legionella* and associated bacterial populations in a hospital hot water system. *System Appl Microbiol.* 38: 198–205.

Benzanson G, Furbridge S, Haldane D, Marrie T. 1992. *In situ* colonization of polyvinyl chloride, brass, and copper by *Legionella pneumophila*. *Can J Microbiol.* 38: 328–330.

Bollin GE, Plouffe JF, Para MF, Hackman B. 1985. Aerosols containing *Legionella pneumophila* generated by shower heads and hot-water faucets. *Appl Environ Microbiol.* 50: 1128–1131.

Bouyer S, Imbert C, Rodier M-H, Yann H. 2007. Long-term survival of *Legionella pneumophila* associated with *Acanthamoeba castellanii* vesicles. *Environ. Microbiol.* 9: 1341–1344. https://doi.org/10.1111/j.1462-2920.2006.01229.x

Brieland JK, Fantone JC, Remick DG, LeGendre M, McClain M, Engleberg NC. 1997. The role of *Legionella pneumophila*-infected *Hartmanella vermiformis* as an infectious particle in a murine model of Legionnaires' disease. *Infect Immun.* 65: 5330–5333.

Brown CM, Nuorti PJ, Breiman RF, Hathcock AL, Fields BS, Lipman HB, Llewellny GC, Hofmann J, Cetron M. 1999. A community outbreak of Legionnaires' disease linked to hospital cooling towers: An epidemiological method to calculate dose of exposure. *Int J Epidemiol.* 28: 353–359.

Buse HY, Morris BJ, Struewing IT, Szabo JG. 2019. Chlorine and monochloramine disinfection of *Legionella pneumophila* colonizing copper and polyvinvy chloride drinking water biofilms. *Appl Environ Microbiol.* 85: e02956–18.

Casini B, Buzzigoli A, Cristini ML, Spagnolo Am, Del Guidice P, Brusaferro S, Poscia A, Moscato U, Valentini P, Baggiani A, Privitera G. 2014. Long-term effects of hospital water network disinfection on *Legionella* and other waterborne bacteria in an Italian university hospital. *Infect Control Hosp. Epidemiol.* 35: 293–299.

Centers for Disease Control and Prevention. 2017 National Notifiable Diseases Surveillance System, Annual Tables of Infectious Disease Data. Atlanta, GA. CDC Division of Health Informatics and Surveillance, 2016. https://www.cdc.gov/nndss/infectious-tables.html.

Ciesielski CA, Blaser MJ, Wang W-LL. 1984. Role of stagnation and obstruction of water flow in isolation of *Legionella pneumophila* from hospital plumbing. *Appl Environ Microbiol.* 48: 984–987.

Cloutman-Green E, Barbosa VL, Jimenez D, Wong D, Dunn H, Needham B, Ciric L, Hartley JC. 2019. Controlling *Legionella pneumophila* in water systems at reduced hot water temperatures with copper and silver ionization. *Am. J. Infect. Control.* 47: 761–766.

Cooper IR, Hanlon GW. 2010. Resistance of *Legionella pneumophila* serotype 1 biofilms to chlorine-based disinfection. *J Hosp Infect.* 74: 152–159.

Correia AM, Gonçalves J, Gomes JP. 2016. Probable person-to-person transmission of Legionnaires' disease. *N Engl J Med.* 374: 497–498.

De Filippis P, Mozzetti C, Messina A, D'Alò GL. 2018. Prevalence of *Legionella* in retirement homes and group homes water distribution systems. *Sci Total Environ.* 643: 715–724.

Den Boer JW, Yzerman EPF, Schellekens J, Lettinga KD, Boshuizen HC, Van Steenbergen JE, Bosman A, Van den Hof S, Van Vliet HA, Peeters MF, Van Ketel RJ, Speelman P, Kool JL, Conyn-Van Spaendonck MAE. 2002. A large outbreak of Legionnaires' disease at a flower show, the Netherlands, 1999. *Emerg Infect Dis.* 8: 37–43.

Ditommaso S, Giacomuzzi M, Ricciadi E, Zotti CM. 2016. Cultural and molecular evidence of *Legionella* spp. colonization in dental unit waterlines: Which is the best method for risk assessment? *Int J Environ Res Publ Hlth.* 13: 211.

Donohue MJ, O'Connell K, Vesper SJ, Mistry JH, King D, Kostich M, Pfaller S. 2014. Widespread molecular detection of *Legionella pneumophila* serogroup 1 in cold water taps across the Unites States. *Environ Sci Technol.* 48: 3145–3152.

Farnham A, Alleyne L, Cimini D, Balter S. 2014. Legionnaires' disease incidence and risk factors, New York, USA, 2002–2011. *Emerg Infect Dis.* 20: 1795–1802.

Fitzhenry R, Weiss D, Cimini D, Balter S, Boyd C, Aleyne L, Stewart R, McIntosh N, Econome A, Lin Y, Rubinstein I, Passaretti T, Kidney A, Lapierre P, Kass D, Varma JK. 2017. Legionnaires' Disease outbreaks and cooling towers, New York City, USA. *Emerg Infect Dis.* 23: 1769–1776.

Garrison LE, Kunz JM, Cooley LA, Moore MR, Lucas C, Schrag S, Sarisky J, Whitney CG. 2016. Vital signs: Deficiencies in environmental control identified in outbreaks of Legionnaires' disease—North America, 2000–2014. *Morbid Mortal Wlky Rpt.* 65: 1–9.

Girolamini L, Dormi A, Pelati T, Somaroli P, Montanare D, Costa A, Savelli F, Martelli A, Grottola A, Serpini GF, Cristino S. 2019. Advances in *Legionella* control by a new formulation of hydrogen peroxide and silver salts in a hospital hot water network. *Pathogens.* 8: 209.

Hayes-Phillips D, Bentham R, Ross K, Whitley H. 2019. Factors influencing *Legionella* contamination of domestic household showers. *Pathogens.* 8: 27. doi: 10.3390/pathogens8010027.

Hershey J, Burrus B, Marcussen V, Notter J, Watson K, Wolford R, Shaffner RE III. 1997. Legionnaires' disease associated with a whirlpool spa display, Virginia September-October, 1996. *Morbid Mortal Wkly Rep.* 46: 83–86.

Jonas D, Meyer H-GW, Matthes P. Hartung D, Jahn B, Daschner FD, Jansen B. 2000. Comparative evaluation of three different genotyping methods for investigation of nosocomial outbreaks of Legionnaires' disease in hospitals. *J Clin Microbiol.* 38: 2284–2291.

Juraskova ES, Sedlackova H, Janska J, Holy O, Lalova I, Matouskova I. 2017. *Legionella* spp. in dental unit waterlines. *Bratisl Med J.* 118: 310–314.

Kuchta JM, States SJ, McGlaughlin JE, Overmeyer JH, Wadowsky RM, McNamare AM, Wolford RS, Yee RB. 1985. Enhanced chlorine resistance of tap water-adapted *Legionella pneumophila* as compared with agar medium-passaged strains. *Appl Environ Microbiol.* 50: 21–26.

Kusnetsov J, Torvinen E, Perola O, Nousiainen T, Katila M-L. 2003. Colonization of hospital water systems by Legionellae, mycobacteria and other heterotrophic bacteria potentially hazardous to risk group patients. *Acta Pathol Microbiol Immunol Scand.* 111: 546–556.

Lau HY, Ashbolt NJ. 2009. The role of biofilms and protozoa in *Legionella* pathogenesis: Implications for drinking water. *J Appl Microbiol.* 107: 368–378.

Liu A, Stout JE, Boldin M, Rugh J, Diven WF, Yu VL. 1998. Intermittent use of copper-silver ionization for *Legionella* control in water distribution systems: A potential option in buildings housing individuals at low risk of infection. *Clin Infect Dis.* 26: 138–140.

Lu J, Buse HY, Gomez-Alvarez V, Struweing I, Santo Domingo J, Ashbolt NJ. 2014. Impact of drinking water conditions and copper materials on downstream biofilm microbial communities and *Legionella pneumophila* colonization. *J Appl Microbiol.* 117: 905–918.

Marrie T, Green P, Burbridge S, Bezanson G, Neale S, Hoffman P.S., Haldane D. 1994. Legionellaceae in the potable water of Nova Scotia hospitals and Halifax residences. *Epidemiol. Infect.* 112: 143–150.

Martinelli F, Caruso A, Moschini L, Turano A, Scarcella C, Speziani F. 2000. A comparison of *Legionella pneumophila* occurrence in hot water tanks and instantaneous devices in domestic nosocomial and community environments. *Curr Microbiol.* 41: 374–376.

Mauchline JWS, Araujo R. Wait R, Dowsett AB, Dennis PJ, Keevil CW. 1992. Physiology and morphology of *Legionella pneumophila* in continuous culture at low oxygen concentration. *J gen Microbiol.* 138: 2371–2380.

Miyamoto M, Yamaguchi Y, Sasatsu M. 2000. Disinfectant effects of hot water, ultraviolet light, silver ions, and chlorine on strains of *Legionella* and nontuberculous mycobacteria. *Microbios.* 101: 7–13.

Moore MR, Pryor M, Fields B, Lucas C, Phelan M, Besser RE. 2006. Introduction of monochloramine into a municipal water system: Impact on colonization of buildings by *Legionella* spp. *Appl Environ Microbiol.* 72: 378–383.

Olsen JS, Aarskaug T, Thrane I, Pourcel C, Ask E, Johansen G, Waagen V, Blatny JM. 2010. Alternative routes for dissemination of *Legionella pneumophila* causing three outbreaks in Norway. *Environ Sci Technol.* 44: 8712–8717.

Pankhurst CL, Coulter W, Philpott-Howard JJ, Harrison T, Warburton F, Platt S, Surman S, Challacombe S. 2003. Prevalence of legionella waterline contamination and *Legionella pneumophila* antibodies in general dental practitioners in London and rural Northern Ireland. *Brit Dent J.* 195: 591–594.

Paszko-Kolva C, Shahamat M, Colwell RR. 1992. Long-term survival of *Legionella pneumophila* serogroup 1 under low-nutrient conditions and associated morphological changes. *FEMS Microbiol. Ecol.* 11: 45–55. https://doi.org/10.1111/j.1574-6968.1992.tb05794.x

Pedro-Botet ML, Stout JE, Yu VL. 2002. Legionnaires' disease contracted from patient homes: The coming of a third plague? *Eur J Clin Microbiol Infect Dis.* 21: 699–705.

Pryor M, Springthorpe S, Riffard S, Brooks T, Huo Y, Davis G, Sattar SA. 2004. Investigation of opportunistic pathogens in municipal drinking water under different supply and treatment regimens. *Water Sci Technol.* 50: 83–90.

Pierre D, Baron JL, Ma X, Sidari FP III, Wagener MM, Stout JE. 2019. Water quality as a predictor of *Legionella* positivity of building water systems. *Pathogens.* 8: 295.

Pourcel C, Visca P, Afshar B, D'Arezzo S, Vergnaud G, Fry NK. 2007. Identification of variable-number tandem-repeat (VNTR) sequences in *Legionella pneumophila* and development of an optimized multiple-locus VNTM analysis typing scheme. *J Clin Microbiol.* 45: 1190–1199.

Ricci ML, Fontana S, Pinci F, Fiumana E, Pedna MF, Farolfi P, Sabattini MAB, Scaturro M. 2012. Penumonia associated with dental unit waterline. *Lancet.* 379: 684.

Rodríguez-Martínez S, Sharaby Y, Pecellín M, Brettar I, Höfle M, Halpern M. 2015. Spatial distribution of *Legionella pneumophila* MLVA-genotypes in a drinking water system. *Water Res.* 77: 119–132.

Rogers J, Dowsett AB, Dennis PJ, Lee JV, Keevil CW. 1994a. Influence plumbing materials on biofilm formation and growth of *Legionella pneumophila* in potable water systems. *Appl Environ Microbiol.* 60: 1842–1851.

Rogers J, Dowsett AB, Dennis PJ, Lee JV, Keevil CW. 1994b. Influence of temperature and plumbing material selection on biofilm formation and growth of *Legionella pneumophila* in a model potable water system containing complex microbial flora. *Appl Environ Microbiol.* 60: 1585–1592.

Saby S, Vidal A, Suty H. 2005. Resistance of Legionella to disinfection in hot water distribution systems. *Water Sci Technol.* 52: 15–28.

Sânchez-Busó L, Comas I, Jorques G, Gonzâlez-Candelas F. 2014. Recombination drives genome evolution in outbreak-related *Legionella pneumophila* isolates. *Nature Genetics.* 46: 1205–1211.

Sânchez-Busó L, Guiral S, Crespi S, Moya V, Camaró ML, Olmos MP, Adriân F, Morera V, Gonzâlez-Morân F, Vanaclocha H, Gonzâlez-Candelas F. 2016. Genomic investigation of a legionellosis outbreak in a persistently colonized hotel. *Front Microbiol.* 6: 1556.

Scaturro M, Buffoni M, Girolamo A, Cristino S, Girolamini L, Mazzotta M, Sabattini MAB, Zaccaro CM, Chetti L, Microbiology Arpa Novara

Laboratory, Bella A, Rota MC, Ricci ML. 2020. Performance of Legiolert® test vs. ISO 11731 to confirm *Legionella pneumophila* Contamination in potable water samples. *Pathogens*. 9: 690. doi: doi:10.3390/pathogens9090690.

Schofield GM, Locci R. 1985. Colonization of components of a model hot water system by *Legionella pneumophila*. *J appl Bacteriol*. 58: 151–162.

Schwake DO, Alum A, Abbaszsdegan M. 2015. Impact of environmental factors on *Legionella* population in drinking water. *Pathogens*. 4: 269–282.

Simmering JE, Polgreen LA, Hornick DB, Sewell DK, Polgreen PM. 2017. Weather-dependent risk for Legionnaires' disease, United States. *Emerg Infect Dis*. 23: 1843–1851.

Singh T, Coogan MM. 2005. Isolation of pathogenic *Legionella* species and legionella-laden amoebae in dental unit waterlines. *J Hosp Infect*. 61: 257–262.

Smith AF, Huss A, Dorevitch S, Heijnen L, Arntzen VH, Davies M, Robert-Du Ry van Beest Holle M, Fujita Y, Verschoor AM, Raterman B, Oesterholt F, Heedrik D, Medema G. 2019. Multiple sources of the outbreak of Legionnaires' disease in Genesee County, Michigan, in 2014 and 2015. *Environ Hlth Persp*. 127: 1–11.

Straus WL, Plouffe JF, File TM Jr, Lipman HB, Hackman BH, Salstrom S-J, Benson RF, Breiman RF. 1996. Risk factors for domestic acquisition of Legionnaires Disease. *Arch Intern Med*. 156: 1685–1692.

Todaro M, Mariotti T, Bisordi C, De Vita E, Valentini P, Costa AL, Casini B, Privitera G, Baggiani A. 2020. Evaluation of *Legionella pneumophila* decrease in hot water network of four hospital buildings after installation of electron time flow taps. *Water*. 12: 210.

Wadowsky RM, Wolford R, McNamara M, Yee RB. 1985. Effect of temperature, pH, and oxygen level on the multiplication of naturally occurring *Legionella pneumophila* in potable water. *Appl Environ Microbiol*. 49: 1197–1205.

Whiley H, Keegan A, Fallowfield H, Bentham R. 2014. Detection of *Legionella, L. pneumophila* and *Mycobacterium avium* complex (MAC) along potable water distribution pipelines. *Int J Environ Res Public Hlth*. 11: 7393–7405.

Woo AH, Yu VL, Goetz A. 1986. Potential in-hospital modes of transmission of *Legionella pneumophila*. Demonstration experiments for dissemination by showers, humidifiers, and rinsing of ventilation bag apparatus. *Amer. J. Med*. 80: 567–573. doi: https://doi.org/10.1016/0002-9343(86)90809-0.

4.2 *Pseudomonas aeruginosa*

4.2.1 Introduction

Pseudomonas aeruginosa is a Gram-negative bacterium that has been recovered from a wide variety of natural and human-engineered environments, including soils, natural waters, and drinking waters (Pellett et al., 1983; Legnani et al., 1999; Diaz et al., 2019). It is also recovered from water and instruments in hospitals that can come in contact with patients (Bédard et al., 2016; Charron et al., 2015; Paline et al., 2017; Rogues et al., 2007). It is an opportunistic pathogen of humans, animals, and plants. *P. aeruginosa* infections include life-threatening dermal infections in burn patients and pneumonia, bacteremia, and urinary tract infections in patients with compromised immune systems (Govan et al., 2007). Mechanical-assisted ventilation is a risk factor for hospital-acquired pneumonia. *P. aeruginosa* is also a major pathogen of cystic fibrosis patients, causing life-threatening pneumonia (Govan et al., 2007; Bianconi et al., 2019; Wargo, 2019). A major challenge to contemporary infection control schemes by *P. aeruginosa* presence in hospitals is its combination in drinking water, and disinfectant and detergent resistance with the production of multiple virulence factors (Agnoli et al., 2012; Grobe et al., 2001; Hota et al., 2009; Matz et al., 2008; Strateva and Yordanov, 2009; Wendel et al., 2015).

A striking feature of *P. aeruginosa* is its resistance to disinfectants, detergents, and a variety of antimicrobial compounds (Strateva and Yordanov, 2009). As such, *P. aeruginosa* presents a difficult problem for control or decontamination; simply *P. aeruginosa* survives most disinfectant and cleaning solutions. Further, there have been reports of *P. aeruginosa* in disinfectant solutions and skin-sterilizing solutions (Favero et al., 1971; Kayser et al., 1975; Weber et al., 2007). As those are employed to clean surfaces or disinfect the skin before injections or surgery, *P. aeruginosa* infections in patients undergoing surgical procedures have been traced to contaminated skin-sterilizing solutions (Weber et al., 2007). Based on the bacterium's characteristics, it should not be surprising that *P. aeruginosa* skin infection (folliculitis) is associated with inadequately disinfected hot tubs or spas (Sausker et al., 1978). There is also a report of *P. aeruginosa* infections amongst infants in a neonatal care unit traced to a water-bath used to thaw frozen plasma for treating the neonates (Muyldermans et al., 1998).

4.2.2 Epidemiology, Diseases, and Risk Factors

P. aeruginosa is a cause of hospital-acquired as well as community-associated life-threatening pneumonia. The U.S. Centers for Disease Control and Prevention (CDC) reported 32,600 cases and 2,700 deaths in 2017, due to multidrug-resistant *P. aeruginosa* (CDC, 2019). *P. aeruginosa*-associated infections are quite severe and require immediate attention, particularly pneumonia (Fujitani et al., 2011). In a meta-analysis of *P. aeruginosa*-associated bronchiectasis including 3,683 patients, *P. aeruginosa* led to increased hospital admissions, severity of disease symptoms, and three-fold increase in mortality (Finch et al., 2015).

4.2.3 Sources and Ecology

P. aeruginosa has been detected in a variety of habitats that it shares with humans, including rivers and lakes (Pellett et al., 1983), rainwater (Kobayashi et al, 2014), drinking water (Reuter et al., 2002), swimming pools (Seyfried and Fraser, 1980), faucet aerator (Cross et al., 1966), bottled mineral waters (Legnani et al., 1999), holy water (Rees and Allen, 1996), humidifiers (Grieble et al., 1970), whirlpool baths (Sausker et al., 1978), hospital water baths (Muyldermans et al., 1998), hospital sinks and drains (Davis et al., 2015; Wendel et al., 2015), and distilled water (Favero et al., 1971).

A variety of techniques have been used to fingerprint and thereby trace *P. aeruginosa* isolates in outbreaks. Pulsed-field gel electrophoresis (PFGE) and repetitive sequence-based PCR (rep-PCR) were used to fingerprint and track metallo-β-lactamase-producing, carbapenem-resistant hospital isolates of *P. aeruginosa* (Wendel et al., 2015). Whole-genome sequencing has been used to track a *P. aeruginosa* outbreak in a neonatal intensive care unit (Davis et al., 2015).

P. aeruginosa and *L. pneumophila* adhered and formed biofilms on normal flora biofilms established on an ethylene-propylene-diene-monomer (EDPM) rubber and on silane cross-linked polyethylene (PE-X) (Moritz et al., 2010). However, unlike *L. pneumophila*, *P. aeruginosa* did not adhere or form biofilms on copper (Moritz et al., 2010). There was little difference in counts/cm^2 over the period of 1, 14, and 28 days incubation (Moritz et al., 2010). However, in those experiments, coupons were incubated in the presence of both microorganisms, thus suspended cells were growing and able to adhere to the coupons throughout the entire course of the incubation (28 days). FISH yielded higher numbers of cells of both *P. aeruginosa* and *L. pneumophila* suggesting that the adhering cells of both were in the viable, but non-culturable state (VBNC), due to copper exposure (Bédard et al., 2014; Moritz et al., 2010).

P. aeruginosa is relatively resistant to antibiotics and disinfectants. A variety of mechanisms contribute to antibiotic resistance in *P. aeruginosa* including (a) antibiotic-inactivating enzymes, (b) reduced outer membrane permeability, and (c) overexpression of efflux systems with a wide range of substrate preferences (Strateva and Yordanov, 2009). Some of those mechanisms (e.g., reduced outer membrane permeability) may contribute to disinfectant resistance. As with other OPPPs, *P. aeruginosa* is resistant to disinfectants used in water treatment such as chlorine (Bédard et al., 2014). Exposure of *P. aeruginosa* to free chlorine (2 mg/L) or copper ions (0.25 mg/L) for 24 h led to a loss of culturability (Bédard et al., 2014). Cells exposed to chlorine lost both viability and culturability, but copper-exposed cells did not lose in viability, only culturability, leading to a VBNC state. Chlorine-exposed cells regained viability (3 h) and culturability (24 h) after depletion of chlorine, while copper-exposed cells regained culturability immediately are following copper removal by chelation (Bédard et al., 2014). Such data forces a re-thinking of standard measurements of chlorine-disinfection efficacy, as it is possible that *P. aeruginosa* cells in a water treatment plant will only be transiently inviable and can be revived following their movement to an environment without chlorine.

Cells of *P. aeruginosa* in biofilms, like other biofilm inhabitants, survive higher dosages of disinfectants, in part due to the presence of layers of cells and extracellular matrix material in the biofilm (Moritz et al., 2010; Shih and Lin, 2010). *P. aeruginosa* can survive periods of stagnation in a water system (e.g., plumbing), due to its ability to utilize nitrate (NO_3^-) as a terminal electron acceptor in the absence of oxygen (Palmer et al., 2007; Sias et al., 1980). In drinking water systems disinfected with chloramine, stagnation may result in *P. aeruginosa* proliferation due, in part, to nitrate-driven respiration (Palmer et al., 2007; Sias et al., 1980).

4.2.4 Transmission

Aerosol transmission of *P. aeruginosa* at a cystic fibrosis patient ward was documented by demonstration of isolates from patients and air samples had the same PFGE patterns (Jones et al., 2003). Using pyocin-typing, a fine-particle humidifier was identified as the source of aerosol transmission of *P. aeruginosa* in a respiratory disease unit (Grieble et al., 1970).

Infections have been linked to *P. aeruginosa* in solutions used for surface sterilization, bronchoscopes washed with non-sterile water or disinfectant solutions, and in-dwelling catheters (Favero et al., 1971; Kayser et al., 1975; Weber et al., 2007). Although most commentators discuss the transfer of drinking water pathogens from water to patient, there is evidence of the patient introduction of *P. aeruginosa* into water taps (Reuter et al., 2002). Alternatively, tap water to patient transmission has been documented (Rogues et al., 2007). *P. aeruginosa* readily forms biofilms (Moritz et al., 2010; Shih and Lin, 2010), so it can persist and grow in flowing systems (e.g., pipes).

One means to reduce exposure of susceptible individuals to *P. aeruginosa* is water filtration. Following detection of an increased frequency of *P. aeruginosa* bacteremia in a hematology unit and documentation that *P. aeruginosa* was present in a variety of possible sources (e.g., water taps, showers, and traps), 0.2 μm pore size filters were installed on shower heads and water taps. Following the installation of the filters, the frequency of blood cultures with *P. aeruginosa* dropped significantly (Vianelli et al., 2006).

4.2.5 Physiological Ecology

The habitats of *P. aeruginosa* overlap with those of humans and their engineered water systems. *P. aeruginosa* is a normal inhabitant of natural and drinking water and has been shown to grow in bottled water (Tamagnini and González et al., 1997), mineral water (Legnani et al., 1999), and even distilled (Favero et al., 1971) and ultrapure water (Kayser et al., 1975). The growth of isolates of *P. aeruginosa* recovered from estuaries, households, and medical clinics in a variety of different laboratory media were the same, indicating that there had been no selection and preference for the growth media composition (Diaz et al., 2018), suggesting that *P. aeruginosa* can survive and grow in a variety of environments via non-mutational adaptation. The variety and scope of its mechanisms for survival in the face of antimicrobials (Strateva and Yordanov, 2009), means there are a few natural or engineered water habitats in which it cannot survive and proliferate.

It has been pointed out that a number of conditions in premise plumbing, namely high and low temperatures, exposure to antimicrobial agents (i.e., chlorine in plumbing and antibiotics in the lung), surfaces for adherence, and high and low oxygen levels, are also found in the cystic fibrosis lung (Wargo, 2019). Thus, the persistence and growth of *P. aeruginosa* in premise plumbing are not inconsistent with infection of the human lung.

There have been reports of *P. aeruginosa* in disinfectant solutions and skin-sterilizing solutions (Kayser et al., 1975; Weber et al., 2007). As those are employed to clean surfaces or to disinfect the skin before injections or surgery, *P. aeruginosa* infections in patients undergoing surgical procedures have been traced to contaminated skin-sterilizing solutions.

P. aeruginosa was one of the first studied for biofilm formation and serves as a model for most investigations of biofilm formation by other microorganisms; I have followed that lead in our investigations of the mycobacteria. A major advantage of biofilm formation is that it prevents washout in flowing systems (e.g., rivers, distribution systems, and premise plumbing). Once established, a biofilm will persist. The persistence in water distribution systems and premise plumbing is supported by the ability of *P. aeruginosa* to grow on low levels of organic carbon (above) and under low oxygen or anaerobic conditions, due to its ability to utilize nitrate as a terminal electron acceptor.

Long-term persistence of *P. aeruginosa* in distribution systems and premise plumbing is supported by its belonging to the amoebae-resisting microorganisms (ARM) (Thomas et al., 2006; 2008). Under conditions of high nutrient, *P. aeruginosa* can grow in protozoa and amoeba (Anderson et al., 2018).

A recent article (Mukherjee et al., 2019) has demonstrated that a light-receptor in *P. aeruginosa* is a kinase and in the presence of light represses biofilm formation and virulence. It is not known whether this activity is present in its close relative, *Burkholderia cepacia*, or other OPPPs, but adds a further dimension (i.e., light) to factors regulating the behavior of OPPPs in water. At the very least, this discovery suggests that biofilm formation in premise plumbing could be reduced by illumination, now we need to devise a method for illumination.

Finally, *P. aeruginosa* produces a range of antimicrobial compounds (including cyanide and phenazines) that modify redox state, act as cell signaling agents, enhance biofilm formation, and increase survival under stress (Pierson and Pierson, 2010). In addition, the antimicrobial properties of some of the secondary metabolites can affect their surrounding microflora (Pierson and Pierson, 2010).

4.2.6 Disinfection of *Pseudomonas aeruginosa*

Because of the relative chlorine-resistance of *P. aeruginosa* (Bédard et al., 2014; 2015; Lee et al., 2011); Seyfried and Fraser, 1980), other methods of disinfection have been assessed. Copper-silver ionization was shown to be able to kill a substantial proportion of planktonic or biofilm-associated cells of *P. aeruginosa, S. maltophilia,* and *A. baumannii* (Shih and Lin, 2010). In spite of 99.99% killing, there was regrowth (i.e., 2- to 3-log increase in number) of *P. aeruginosa* cells after 24 h. No substantial regrowth of either *S. maltophilia* or *A. baumannii* was noted (Shih and Lin, 2010).

4.2.7 Genome Variation and Population Diversity

Genome reorganization in *P. aeruginosa* may be a hallmark of OPPPs. It is adaptive in the sense that the generation of phenotypic variants allows the survival of the population in spite of unexpected stresses. Some stresses can be expected by an environmental microorganism, such as exposure to UV-irradiation, desiccation, and toxic oxygen radicals. Although the driver(s) for phenotypic and genotypic variability has not yet been described, it has been hypothesized that the high frequency of genetic variants in *P. aeruginosa* in tissue from chronically-infected patients (e.g., cystic fibrosis) is due to both intrinsic and environmental mutagenesis (Rodriguez-Rojas et al., 2012). Two approaches have been employed to characterize genome evolution of *P. aeruginosa,* namely studies of *P. aeruginosa* strains from a single patient over time (Bianconi et al., 2019) or studies of large numbers of *P. aeruginosa* strains in patients and the environment (Kidd et al., 2012; Rodriguez-Rojas et al., 2012)

From a single patient, whole-genome sequences were obtained from 40 *P. aeruginosa* isolates obtained over an 8-year period (2007–2014). Mutations were associated with the appearance of antibiotic-resistant isolates amongst different clonal descendants of the original, infecting strain (Bianconi et al., 2019). Further genome analysis indicated that the driver of the population variability was due to horizontal gene transfer and/or convergent evolution (Bianconi et al., 2019).

In a study of human and environmental isolates of *P. aeruginosa* in Queensland, Australia, genetic diversity could be ascribed to frequent recombination (Kidd et al., 2012). Further, *P. aeruginosa* strains from patients were non-clonal with most representing a random sample of the overall environmental and human-associated population, and that there was little association between genotype and environment (Kidd et al., 2012). As noted for *A. baumannii* and *S. maltophilia, P. aeruginosa* exhibits an inherently high rate of genome reorganization that serves as the substrate for selection in a variety of different environments.

4.2.8 *Pseudomonas aeruginosa* as an Indicator for OPPPs

It has been well-documented that counts of neither *Escherichia coli* and coliforms, nor the heterotrophic plate count (HPC) bacteria correlate or are surrogates for the presence of opportunistic premise plumbing pathogens in drinking water (Allen et al., 2004; Bartram et al., 2004; Briancesco et al., 2014; Kobyashi et al., 2014; Gensberger et al., 2015; Hamilton et al., 2016). It is likely that the lack of correlation between numbers of those two measurements is due, in part, to the differences in the physiologic ecology of OPPPs and fecal-origin bacteria. As documented throughout this book, OPPPs are resistant to disinfectants, grow in drinking water, readily form biofilms, and are more resistant to conditions of stress (e.g., high temperature, desiccation, and presence of reactive oxygen metabolites) than are *E. coli*, coliforms, and HPC bacteria. The author (Falkinham, 2017) and others (de Victoria and Galván, 2001) have proposed that *P. aeruginosa* be considered an indicator for drinking water quality. *P. aeruginosa* is an appealing indicator as it grows rapidly and selective media are available for its isolation, detection, and enumeration, such as *Pseudomonas* Isolation Agar (Becton Dickenson, Sparks, Maryland, USA) and Pseudalert®/Quanti-Tray® (IDEXX Laboratories, Westbrook, Maine, USA; Sartory et al., 2015). Much more study is needed to validate such a measure, especially in light of the observation that waters from treatment plants, distribution systems, and premise plumbing all have unique microbiomes (Dias et al., 2019).

4.2.9 Notes on the *Burkholderia cepacia* Complex

4.2.9.1 Introduction

Rather than dedicate a whole chapter on the *Burkholderia cepacia* complex, I will rely upon the similarity of behaviors of *Pseudomonas aeruginosa* and the *Burkholderia cepacia* complex. In this section, I will restrict the text to presenting unique features of members of the *B. cepacia* complex.

First, the *B. cepacia* complex includes the following species: *B. cepacia, B. cenocepacia, B. multivorans, B. stabilis, B. vietnamiensis,*

B. dolosa, B. ambifaria, B. anthina, and *B. pyrrocinia*. Isolates of *B. cenocepacia* account for approximately 45% of *B. cepacia* complex infections in individuals with cystic fibrosis in the United States, but up to 80% in cystic fibrosis patients in Canada and some European countries (Lucero et al., 2010). One feature that distinguishes *B. cepacia* from *P. aeruginosa* is the presence of three chromosomes in *B. cepacia*, one of which appears to be a virulence plasmid (Agnoli et al., 2012).

4.2.9.2 *Burkholderia cepacia* infection and transmissibility

Although both *P. aeruginosa* and *B. cepacia* complex infects the lungs of cystic fibrosis patients, infection with *B. cepacia* is more invasive (Govan et al., 2007). Further, there is evidence of person-to-person (Lipuma et al., 1990) and strain-dependent transmission of *B. cepacia* complex strains between cystic fibrosis patients (Govan et al., 2007). There are two known epidemic and virulence markers, namely cable pili and *B. cepacia* epidemic strain marker (BCESM), and their presence suggests that strains with either or both characteristics would be more invasive and transmissible (Clode et al., 2000). Like *P. aeruginosa*, nosocomial infections caused by *B. cepacia* have been linked to their presence in hospital water (Nasser et al., 2004) and ventilators (Guo et al., 2017). Using pulsed-field gel electrophoresis (PFGE), *B. cepacia* complex infections in ventilated cystic fibrosis patients were linked to isolates from sinks in the patients' intensive care unit (Lucero et al., 2014).

4.2.9.3 *Burkholderia cepacia* ecology

B. cepacia complex isolates have been recovered from a variety of natural and engineered water systems, including hospital premise plumbing (Nasser et al., 2004). *B. cepacia* complex species are waterborne, but isolates have been shown to survive on surfaces (Drabick et al., 1996), suggesting they are desiccation-resistant. Although rare amongst the OPPPs, *B. cepacia* complex species have been detected in wastewaters (Merk et al., 2001).

Members of the genus *Burkholderia* are the producers of a wide variety of antimicrobial compounds (Chiarini et al., 2006). A secondary metabolite from an isolate of *B. ambifaria* was shown to produce the anti-fungal burkholdines (Falkinham et al., 2014).

References

Agnoli K, Schwager S, Uehlinger S, Vergunst A, Viteri DF, Nguyen DT, Sokol PA, Carlier A, Eberl L. 2012. Exposing the third chromosome of *Burkholderia cepacia* complex strains as a virulence plasmid. *Mol Microbiol.* 83: 362–278.

Allen MJ, Edberg SC, Reasoner DJ. Heterotrophic plate count bacteria—what is their significance in drinking water? *Int. J. Food Microbiol.* 2004; 92: 265–274.

Andersson A, Alinder J, Mathisen P, Hägglund J, Bäckman S, Nilsson E, Sjökin A, Thelaus J. 2018. Predators and nutrient availability favor-protozoa-resisting bacteria in aquatic systems. *Sci. Rpts.* 8: 8415.

Bartram J, Cotruvo J, Exner M, Fricker C, Glasmacher A. 2004. Heterotrophic plate count measurement in drinking water safety management. *Int. J. Food Microbiol.* 92: 241–247.

Bédard E, Charron D, Lalancette C, Déziel, Prévost M. 2014. Recovery of *Pseudomonas aeruginosa* culturability following copper- and chlorine-induced stress. *FEMS Microbiol Lttr.* 356: 226–234.

Bédard E, Laferrièrre C, Charron D, Lalancette C, Renaud C, Desmarais N, Déziel E, Prévost M. 2015. Post-outbreak investigation of *Pseudomonas aeruginosa* faucet contamination by quantitative polymerase chain reaction and environmental factors affecting positivity. *Infect Contrl Hosp Epidemiol.* 36: 1337–1343.

Bédard E, Prévost M, Déziel E. 2016. *Pseudomonas aeruginosa* in premise plumbing of large buildings. *Microbiol Open.* 00: 1–20. doi: 10.1002/mbo3.391.

Bianconi I, D'Arcangelo S, Esposito A, Benedet M. Fiffer E, Dinnella G, Gualdi P, Schinella M, Baldo E, Donati C, Jousson O. 2019. Persistence and microevolution of *Pseudomonas aeruginosa* in the cystic fibrosis lung: A single-patient longitudinal genomic study. *Front Microbiol.* 9: 3242.

Briancesco R, Alaimo C, Bonanni E, Delle Site A, Di Gianfilippo F, Grassano L, Moscatelli R, Ottaviano C, Paradiso R, Quintiliani S, Semproni M, Bonadonna L. 2014. An Italian investigation on non-tuberculous mycobacteria in an urban water supply. *Ann Ig.* 26: 264–271.

CDC. 2019. Multidrug-resistant *Pseudomonas aeruginosa.* https://www.cdc.gov/drugresistance/pdf/threats-report/pseudomonas-aeruginosa-508.pdf

Charron D, Bédard E, Lalancette C, Laferrière C, Prévost M. 2015. Impact of electronic faucets and water quality on the occurrence of *Pseudomonas aeruginosa* in water: A multi-hospital study. *Infect Contrl Hosp Epidemiol.* 36: 311–319.

Chiarini L, Bevivino A, Dalmastri C. 2006. *Burkholderia cepacia* complex species: Health hazards and biotechnological potential. *Trends Microbiol.* 14: 277–286.

Clode FE, Kaufmann MF, Malnick H. 2000. Distribution of genes encoding putative transmissibility factors among epidemic and nonepidemic strains of *Burkholderia cepacia* from cystic fibrosis patients in the United Kingdom. *J Clin Microbiol.* 38: 1763–1766.

Cross DF, Benchimol A, Dimond EG. 1966. The faucet aerator: A source of *Pseudomonas* infection. *N Eng J Med.* 274: 1430–1431.

Davis RJ, Jenson SO, Van Hal S, Espedido B, Gordon A, Farhat R, Chan R. 2015. Whole-genome sequencing in real-time investigation and management of a *Pseudomonas aeruginosa* outbreak on a neonatal intensive care unit. *Infect Contr Hosp Epidemiol.* doi: 10.1017/ice.2015.133.

de Victorica J, Galván M. 2001. *Pseudomonas aeruginosa* as an indicator of health risk in water for human consumption. *Water Sci Technol.* 43: 49–52.

Dias VCF, Durand A-A, Constant P, Prévost M, Bédard E. 2019. Identification of factors affecting bacterial abundance and community structures in a full-scale chlorinated drinking water distribution system. *Water.* 11: 627. doi: 10.3390/w11030627.

Diaz KE, Remold SK, Onyiri O, Bozeman M, Raymond PA, Turner PE. 2018. Generalized growth of estuarine, household and clinical isolates of *Pseudomonas aeruginosa*. *Front Microbiol.* 9: 305.

Drabick JA, Gracely EJ, Heidecker GJ. 1996. Survival of *Burkholderia cepacia* on environmental surfaces. *J Hosp Infect.* 32: 267–276.

Falkinham JO III. 2017. *Pseudomonas aeruginosa* as an indicator for opportunistic premise plumbing pathogens. *JSM Microbiol.* 5: 1042.

Favero MS, Carson LA, Bond WW, Petersen NJ. 1971. *Pseudomonas aeruginosa*: Growth in distilled water from hospitals. *Science.* 175: 836–838.

Finch S, McDonnell MJ, Abo-Leyah H, Aliberti S, Chalmers JD. 2015. A comprehensive analysis of the impact of *Pseudomonas aeruginosa* colonization on prognosis in adult bronchiectasis. *Ann Am Thor Soc.* 12: 1602–1611.

Fujitani S, Sun H-Y, Yu VL, Weingarten JA. 2011. Pneumonia due to *Pseudomonas aeruginosa*. Part I: Epidemiology, clinical diagnosis, and source. *Chest.* 139: 909–919.

Gensberger ET, Gössl E-M, Antonelli L, Sessitsch A, Kostić T. 2015. Effect of different heterotrophic plate count method on the estimation of the composition of the culturable microbial community. *Peer J.* 3: e862. doi 10.7717/peerj.862.

Govan JRW, Brown AR, Jones AM. 2007. Evolving epidemiology of *Pseudomonas aeruginosa* and the *Burkholderia cepacia* complex in cystic fibrosis lung infection. *Future Med.* 2.

Grieble HG, Colton FR, Bird TJ, Toigo A, Criffith LG. 1970. Fine-particle humidifiers. *New Engl J Med.* 282: 531–535.

Grobe S, Wingnder J, Flemmin H-K. 2001. Capability of mucoid *Pseudomonas aeruginosa* to survive in chlorinated water. *Int J Hyg Environ Hlth.* 139–142.

Guo LP, Li G, Wang J, Zhao X, Wang S, Zhai L, Jia H, Cao B. 2017. Suspicious outbreak of ventilator-associated pneumonia caused by *Burkholderia cepacia* in a surgical intensive care unit. *Am J Infect Control.* 45: 660–666.

Hamilton KA, Ahmed W, Palmer A, Sidhu JPS, Hodgers L, Toze S, Haas CN. 2016. Public health implications of *Acanthamoeba* and multiple potential opportunistic pathogens in roof-harvested rainwater tanks. *Environ Res.* 150: 320–327.

Hota S, Hirji Z, Stockton K, Lemieux C, Dedier H, Wolfaardt G, Gardam MA. 2009. Outbreak of multidrug-resistant *Pseudomonas aeruginosa* colonization and infection secondary to imperfect intensive care unit room design. *Infect Contrl Hosp Epidemiol.* 30: 25–33.

Jones AM, Govan JRW, Doherty CJ, Dood ME, Isalska BJ, Stanbridge TN, Webb AK. 2003. Identification of airborne dissemination of epidemic multi-resistant strains of *Pseudomonas aeruginosa* at a CF center during a cross infection outbreak. *Thorax.* 58: 525–527.

Kayser WV, Hickman KCD, Bond WW, Favero MS, Carson LA. 1975. Bacteriological evaluation of an ultra-pure water-distilling system. *Appl Microbiol.* 30: 704–706.

Kidd TJ, Ritchie SR, Ramsay KA, Grimwood K, Bell SC. 2012. *Pseudomonas aeruginosa* exhibits frequent recombination, but only a limited association between genotype and ecological setting. *PLoS ONE.* 7: e44199.

Kobayashi M, Oana K, Kawakami Y. 2014. Incidence of *Legionella* and heterotrophic bacteria in household rainwater tanks in Azumino, Nagano prefecture, Japan. *Microbiol Immunol.* 58: 15–21.

Lee CS, Wetzel K, Buckley T, Wozniak D, Lee J. 2011. Rapid and sensitive detection of *Pseudomonas aeruginosa* in chlorinated water and aerosols targeting gyrB gene using real-time PCR. *J Appl Microbiol.* 111: 893–903.

Legnani P, Leoni E, Rapuano S, Turin D, Valenti C. 1999. Survival and growth of *Pseudomonas aeruginosa* in natural mineral water: A 5-year study. *Intl J Food Microbiol.* 53: 153–158.

Lipuma JJ, Dasen SE, Nielson DW, Stern RC, Stull TL. 1990. Person-to-person transmission of *Pseudomonas cepacia* between patients with cystic fibrosis. *Lancet.* 336: 1094–1096.

Loveday HP, Wilson JA, Kerr K, Pitchers R, Walker JT, Browne J. 2014. Association between healthcare water systems and *Pseudomonas aeruginosa* infections: A rapid systematic review. *J Hosp Infect.* 86: 7–15.

Lu J, Struewing I, Vereen E, Kirby AE, Levy K, Moe C, Ashbolt N. 2015. Molecular detection of *Legionella* spp. and their associations with *Mycobacterium* spp., *Pseudomonas aeruginosa* and amoeba hosts in a drinking water distribution system. *J Appl Microbiol.* 120: 509–521.

Lucero CA, Cohen AL, Trevino I, Rupp, AH, Harris M, Forkan-Kelly S, Noble-Wang J, Jensen B, Shams A, Arduino MJ, LiPuma JJ, Gerber SI, Srinivasan A. 2011. Outbreak of *Burkholderia cepacia* complex among ventilated pediatric patients linked to hospital sinks. *Am J Infect Control.* 39: 775–778.

Matz C, Moreno AM, Ahede M, Manefield M. Hauser AR, Givskov M, Kjelleberg S. 2008. *Pseudomonas aeruginosa* uses type III secretion system to kill biofilm-associated amoebae. *Int Soc Microb Ecol J.* 2: 843–852.

Moritz MM, Flemmin H-C, Wingender J. 2010. Integration of *Pseudomonas aeruginosa* and *Legionella pneumophila* in drinking water biofilms grown on domestic plumbing materials. *Int J Hyg Environ Hlth.* 213: 190–197.

Mukherjee S, Jemielita M, Stergioula V, Tikhonov M, Bassler BL. 2019. Photosensing and quorum sensing are integrated to control *Pseudomonas aeruginosa* collective behaviors. *PLoS Biol.* 17: e3000579. doi: 10.1371/journalpbio.3000579.

Muyldermans G, de Smet F, Pierard D, Steenssens L, Stevens D, Bougatef A, Lauwers S. 1998. Neonatal infections with *Pseudomonas aeruginosa* associated with a water-bath used to thaw fresh frozen plasma. *J Hosp Infect.* 39: 309–314.

Nasser RM, Rahi AC, Hadda MF, Daoud Z, Irani-Hakime N, Almawi W. 2004. Outbreak of *Burkholderia cepacia* bacteremia traced to contaminated hospital water used for dilution of an alcohol skin antiseptic. *Infect Control Hosp Epidemiol.* 25: 231–239.

Paling FP, Wolkewitz M, Depuydt P, de Bus L, Sifakis F, Bonten MJM, Kluytmans JAJW. 2017. *P. aeruginosa* colonization at ICU admission as a risk factor for developing *P. aeruginosa* ICU pneumonia. *Antimicrob Resist Infect Contrl.* 6: 38.

Palmer KL, Brown SA, Whitley M. 2007. Membrane-bound nitrate reductase is required for anaerobic growth in cystic fibrosis sputum. *J Bacteriol.* 189: 4449–4455.

Pellett S, Bigley DV, Grimes DJ. 1983. Distribution of *Pseudomonas aeruginosa* in a riverine ecosystem. *Appl Environ Microbiol.* 45: 328–332.

Pierson LS, Pierson EA. 2010. Metabolism and function of phenzaines in bacteria: Impacts on the behavior in the environment and biotechnological processes. *Appl. Microbiol. Biotechnol.* 86: 1659–1670.

Rees JC, Allen KD. 1996. Holy water: A risk factor for hospital-acquired infection. *J Hosp Infect.* 32: 51–55.

Reuter S, Sigge A, Wiedeck H, Trautmann M. 2002. Analysis of transmission pathways of *Pseudomonas aeruginosa* between patients and tap water outlets. *Crit Care Med.* 30: 2222–2228.

Rodriguez-Rojas A, Oliver A, Blázquez J. 2012. Intrinsic and environmental mutagenesis drive diversification and persistence of *Pseudomonas aeruginosa* in chronic lung infection. *J Infect Dis.* 205: 121–127.

Rogues A-M, Boulestreau H, Lashéras A, Boyer A, Gruson D, Merle C, Castaing Y, Bébear CM, Gachie J-P. 2007. Contribution of tap water to patient colonization with *Pseudomonas aeruginosa* in a medical intensive care unit. *J Hosp Infect.* 67: 72–78.

Sartory DP, Pauly D, Garrec N, Bonadonna L, Semproni M, Schell C, Reimann A, Firth SJ, Thom C, Hartemann P, Exner M, Baldauf SJ, Lee S, Lee JV. Evaluation of an MPN test for the rapid enumeration of

Pseudomonas aeruginosa in hospital waters. *KJ Water Health.* 13: 427–435.

Sausker WF, Aeling JL, Fitzpatrick JE, Judson FN. 1978. *Pseudomonas* folliculitis acquired from a health spa whirlpool. *J Am Med Assoc.* 239: 2362–2365.

Seyfried PL, Fraser DJ. 1980. Persistence of *Pseudomonas aeruginosa* in chlorinated swimming pools. *Can J Microbiol.* 26: 350–355.

Shih H-S, Lin YE. 2010. Efficacy of copper-silver ionization in controlling biofilm- and plankton-associated waterborne pathogens. *Appl Environ Microbiol.* 76: 2032–2035.

Sias SR, Stouthamer AH, Ingraham JL. 1980. The assimilatory and dissimilatory nitrate reductases of *Pseudomonas aeruginosa* are encoded by different genes. *J Gen Microbiol.* 118: 229–234.

Strateva T, Yordanov D. 2009. *Pseudomonas aeruginosa*: A phenomenon of bacterial resistance. *J Med Microbiol.* 58: 1133–1148.

Tamagnini LM, González. 1997. Bacteriological stability and growth kinetics of *Pseudomonas aeruginosa* in bottled water. *J Appl Microbiol.* 83: 91–94.

Thomas V, Herrera-Rimann K, Blanc DS, Greub G. 2006. Biodiversity of amoebae and amoebae-resisting bacteria in a hospital water network. *Appl. Environ. Microbiol.* 72: 2428–2438.

Thomas V, Loret J-F, Jousset M, Greub G. 2008. Biodiversity of amoebae and amoebae-resisting bacteria in a drinking water treatment plant. *Environ. Microbiol.* 10: 2728–2745.

Wargo MJ. 2019. Is the potable water system an advantageous preinfection niche for colonizing the cystic fibrosis lung? mBio 10e00883-19. doi:10.1128;mBio.00883-19.

Weber DJ, Rutala WA, Sickbert Bennett EE. 2007. Outbreaks associated with contaminated antiseptics and disinfectants. *Antimicrob. Agents Chemother.* 51: 4217–4224.

Wendel AF, Kolbe-Busch S, Ressina S, Schulz-Röbbecke R, Kindgen-Milles D, Lorenz C, Pfeffer K, MacKenzie CR. 2015. Detection and termination of an extended low-frequency hospital outbreak of GIM-1-producing *Pseudomonas aeruginosa* ST111 in Germany. *Am J Infect Contrl.* 43: 635–639.

Wilson MG, Nelson RC, Phillips LH, Boak RA. 1961. New source of *Pseudomonas aeruginosa* in a nursery. *J Am Med Assoc.* 175: 1146–1148.

4.3 *Mycobacterium avium* Complex (MAC)

4.3.1 Introduction

Little did I realize that in October 2015 when I read an article in *Clinical Infectious Disease* that reported infections caused by *Mycobacterium chimaera* in cardiac surgery patients (Sax et al., 2015) that my life would be dominated by heater-coolers for the next four years. *M. chimaera* is a member of the MAC, a group of environmental opportunistic pathogens whose sources of infections have been shown to include drinking water and soils. The heater-cooler infections were linked to aerosols produced by a particular heater-cooler, the Sorin 3T, and the instruments were colonized by *M. chimaera* in the drinking water at a manufacturing facility in Munich, Germany (Haller et al., 2016).

The members of the MAC are waterborne opportunistic pathogens that cause pulmonary infection, currently, but not exclusively, in slender, taller, and older women and men, cervical lymphadenitis in children, and bacteremia in immunodeficient individuals (Marras and Daley, 2002). Pulmonary infections have been tracked by DNA fingerprinting to showerheads (Falkinham et al., 2008; Tzou et al., 2020) and household plumbing (Falkinham, 2011; Lande et al., 2019). The numbers of *M. avium* increase as drinking water move from the treatment plant to home (Falkinham et al., 2001). The majority of *Mycobacterium* spp. infections in the United States are caused by the MAC and that group serves as a representative of the group of all environmental mycobacteria, called the nontuberculous mycobacteria (NTM). Although NTM disease is not required to be reported, estimates based on hospital admissions show that the prevalence of infection is increasing at a rate of between 8% and 10% annually with approximately 30,000 cases in the United States at present (Marras et al., 2007; Billinger et al., 2009).

4.3.2 The MAC

The MAC currently consists of four subspecies of *M. avium*, namely subsp. *avium*, subsp. *silvaticum*, subsp. *hominissuis*, and subsp. *Paratuberculosis;* and eight species, namely *M. intracellulare,*

M. chimaera, M. colombiense, M. marseillense, M. timonense, M. bouchedurhonense, M. arosiense, and *M. ituriense* (Tortoli, 2003). Within the MAC, *M. avium* subsp. *hominissuis, M. intracellulare,* and *M. chimaera* are the major human pathogens. Only recently, *M. chimaera* has been identified as a human pathogen (Tortoli et al., 2004) and distinguished from *M. intracellulare.* Standard methods for identification, particularly 16S rRNA and *hsp-65* gene-dependent methods are unable to distinguish *M. intracellulare* from *M. chimaera* (Iakhiaeva et al., 2013). Although the process of re-evaluation of the identity of *M. intracellulare* isolates has just been initiated, the initial data suggests that a substantial proportion of *M. intracellulare*-infected patients are infected with *M. chimaera.* Further, it has been shown that *M. intracellulare* is not, as is *M. avium* subsp. *hominissuis,* found in drinking water. Isolates from drinking water and plumbing biofilms have been shown to be exclusively *M. chimaera,* not *M. intracellulare* (Wallace et al., 2013). The ecology of *M. intracellulare* is different from *M. avium* and *M. chimaera,* and it is not found in water and the sources of infection for *M. intracellulare* are different (perhaps soil). The current trend of reporting *M. avium* subsp. *hominissuis, M. chimaera,* or *M. intracellulare* infections as MAC or MAI clearly obscures differences in their distribution, sources, and treatment options.

4.3.3 Diseases, Epidemiology, and Risk Factors

Members of the MAC have been reported to cause pulmonary disease, cervical lymphadenitis in young children with erupting teeth, and bacteremia in immunosuppressed individuals (Marras and Daley, 2002). Strikingly, almost all infections in HIV-infected individuals have been shown due to *M. avium* subsp. *hominissuis* infection (Drake et al., 1988). Guidelines for diagnosis and treatment are available (Griffith et al., 2007).

Based on hospital records, there is evidence of counties in the United States that have significantly higher frequencies of mycobacterial infection (Adjemian et al., 2012). Montgomery County, Pennsylvania, is one example, as is New York City. One reason for the higher frequency of mycobacterial pulmonary disease cases in New York City is the very high numbers of mycobacteria in premise plumbing that I suspect are due to the recirculation of hot water in residential towers (Tichenor et al., 2012). Other hot spots include the bayou-counties of Louisiana, as MAC appears to be estuarine bacteria (George et al., 1980).

Risk factors for MAC pulmonary disease include prior lung damage from infection (e.g., *M. tuberculosis*) or occupational exposure to dusts (e.g., farmers) or silica (e.g., miners) (Marras and Daley, 2002). Another risk factor is immunodeficiency, due to either infection (e.g., HIV) or immunosuppression due to cancer (Redelman and Sepkowitz, 2010), or drug treatment (Simpson et al., 1982). In HIV-infected individuals, *M. avium* numbers can be up to 100,000 CFU/mL with even higher numbers (10,000,000 CFU/gm) in tissue (Torriani et al., 1996). Many of the new species of *Mycobacterium* identified and described in the last 20 years (Tortoli, 2003) have been recovered from HIV-infected patients. Sadly, it appears in those individuals who are sensitive to sensors of NTM in the environment. Risk factors for *M. avium* infection in the HIV-infected include consumption of spring water, consumption of raw fish, and showering outside the home (von Reyn et al., 2002).

Another risk factor for MAC pulmonary disease may be gastric esophageal reflux disease (GERD, Thomson et al., 2007). The majority of MAC pulmonary infections in the United States are found in slender, taller, and older women and men (Prince et al., 1989; Guide and Holland et al., 2002). Some of those women were carriers (i.e., haploid) for mutant alleles leading to cystic fibrosis or α-1-antitrypsin deficiency (Kim et al., 2005).

A proportion of cases of chronic rhinosinusitis is due to NTM infection (Tichenor et al., 2012). To date, there appear to be no predominant species, as both rapidly and slowly growing *Mycobacterium* spp. have been recovered (Tichenor et al., 2012). It is likely that the chronic nature of the infections is due, in part, to the relative resistance of NTM to commonly used, non-anti-mycobacterial antibiotics. In a number of cases of chronic sinusitis, the infections were traced by DNA fingerprinting to household water used for sinus irrigation (Tichenor et al., 2012).

Due to the presence of MAC in drinking water and their relative resistance to high temperature and disinfectants, device-associated pseudo-infections have been identified. These can present as outbreaks of NTM infection associated with the use of bronchoscopes. Insufficient disinfection of bronchoscopes was linked to *M. avium* and *M. intracellulare* (possibly *M. chimaera*) infections, solved by rigorous attention to laboratory procedures for cleaning and disinfection of bronchoscopes (Falkinham, 2010). Long-term use of indwelling catheters has also been linked to *M. avium* infection (Schelonka et al., 1994).

M. chimaera infections were reported in patients following cardiovascular surgery where the Sorin 3T heater-cooler was used (Sax et al., 2015). The infections were traced to the presence of the infecting *M. chimaera* in the water reservoirs of heater-coolers used to control blood temperature in patients during cardiac surgery. Patients throughout the world were shown to be infected by the same strain of *M. chimaera* by genome sequencing and were introduced (inoculated) into the Sorin 3T heater-cooler during final functional testing at the factory in Munich, Germany, before shipping (Haller et al., 2016).

Recently, there have been outbreaks of *Mycobacterium* spp. cervical lymphadenitis linked to dental clinics serving children (Hatzenbuehler et al., 2017; Singh et al., 2018). The children had undergone a "pulpotomy", where the decayed "baby" tooth has been removed without damaging the underlying "adult" tooth. Infection appears within two weeks following surgery. The infections do not resolve spontaneously, but get progressively worse over time with the appearance of an open, draining sinus (Wolinsky, 1995). Surgical excision of the infected cervical lymph node is the most effective treatment; antibiotic therapy is of little help (Wolinsky, 1995). The infecting *Mycobacterium* spp. strains have been linked to their presence in water and biofilms in the "dental unit", the free-standing tower that provides water, suction, and power for dental procedures (Hatzenbuehler et al., 2017; Singh et al., 2018). The source of the *Mycobacterium* spp. has been shown to be the local drinking water supply and inadequate disinfection protocols required to eradicate the disinfectant-resistant *Mycobacterium* spp. led to their proliferation and persistence in the dental unit (Hatzenbuehler et al., 2017; Singh et al., 2018). Disinfectant resistance is increased as a consequence of the adherence of the hydrophobic *Mycobacterium* spp. cells to the surfaces of the "dental unit" where they grow to form biofilms (Steed and Falkinham, 2006).

Diseases caused by MAC and other NTM infections are not required to be reported, as are *L. pneumophila* infections, exact numbers of infections are limited. However, estimates of the number of hospitalized NTM-infected patients and the cost of their hospitalization have been gained. Although MAC numbers are not separated by species, it should be remembered that *M. avium* is the predominant *Mycobacterium* spp. infecting individuals in the United States and Canada. In a study based on data from the Healthcare Cost and Utilization Project (HCUP) of the U.S. Department of Health and Human Services, 20,049 hospital discharges were recorded from 2001 through 2012. The total associated cost of these discharges was $903,767,292 for pulmonary NTM disease (Mirsaeidi et al., 2015). Further, the yearly national hospital costs of pulmonary NTM disease increased significantly from a mean of $42 million in 2001 to $110 million in 2012 (Mirsaeidi et al., 2015). Almost identical estimates for the cost of *Mycobacterium* spp. infections were reached by Collier et al. (2012) in the United States and similar high costs were reported in Europe by Goring et al. (2018).

4.3.4 Transmission Pathways

Transmission pathways for pulmonary disease caused by members of the MAC are primarily through aerosolization. The hydrophobic mycobacterial cells are readily aerosolized from water and, actually, numbers are increased 1,000–10,000-fold in water droplets ejected from water surfaces (Parker et al., 1983). Aerosolization can occur in showers (Falkinham et al., 2008), humidifiers (Hamilton and Falkinham, 2018), hot tubs or spas (Kahana et al., 1997; Katsuda et al., 2018), and heater-cooler instruments used during cardiac surgery (Sax et al., 2015). Pulmonary infection might also be a consequence of gastric reflux disease following ingestion of mycobacterial-laden liquids (Thomson et al., 2007).

Soil dust generated during gardening activities or even weather events appears to be another route of pulmonary infection (De Groote et al., 2006). Potting soils of *M. intracellulare*-infected patients yielded *M. intracellulare* on soil particles that could be inhaled and enter the alveoli (De Groote et al., 2006). Peat-rich potting soils have especially high numbers of *Mycobacterium* spp., 1 million per gram (De Groote et al., 2006).

The route of infection for cervical lymphadenitis in children with erupting teeth is likely through ingestion of MAC-containing soils or water (Wolinsky, 1995) or through squirting water in the mouths of children during tooth removal (Hatzenbuchler et al., 2017; Singh et al., 2018).

4.3.5 Habitats and Sources of the MAC

Members of the MAC are widely distributed in soils and natural and engineered water systems. Literally, humans are surrounded by mycobacteria (Falkinham, 2009). MAC has been isolated from soils, in especially high numbers (1 million per gm) from peat-rich soils (De Groote et al., 2006). Analysis of soil and water samples from the southeastern coasts of the United States from Delaware to Louisiana has shown that MAC is estuarine (Falkinham et al., 1980). MAC is in high numbers in the fresh and brackish water of the Chesapeake and Delaware Bays (Falkinham et al., 1980; Kirschner et al., 1992) and can grow in water containing 2% NaCl, but not in ocean water (George et al., 1980). One reason for its residence in coastal swamps and estuaries is that its growth is stimulated by humic and fulvic acids, the breakdown products of the dark-woody structural constituents of trees and shrubs and major constituents of the dark brown acidic water of the United States coastal estuaries (Kirschner et al., 1995).

Premise plumbing is an ideal habitat for MAC, one created by humans. It provides sufficient nutrient for growth as MAC is oligotrophic (Norton et al., 2005) and it enjoys periodic exposure to elevated temperatures that allow growth while not killing cells (Schultz-Röbbecke and Buchholtz, 1992). Further, periods of stagnation do not limit the MAC growth, as they can grow at 6% oxygen levels (Lewis and Falkinham, 2017). Finally, premise plumbing provides surfaces for adherence and biofilm-formation by the hydrophobic MAC cells (Mullis and Falkinham, 2017) to prevent washout.

4.3.6 Physiological Ecology of the MAC

M. avium grows very slowly (1 generation per day in rich laboratory media), but readily adheres and forms biofilms on pipe surfaces (Mullis and Falkinham, 2017), so it is not washed out. *M. avium* is very disinfectant-resistant (Taylor et al., 2000) and that resistance is increased by acclimation to drinking water (Taylor et al., 2000) and its residence in biofilms (Steed and Falkinham, 2006). *M. avium* is relatively resistant to elevated temperature [90% survive exposure to 50 °C (125 °F) for 60 min (Schulze-Röbbecke and Buchholtz, 1992)], which likely contributes to its residence in households (Falkinham, 2011; Lande et al., 2019) through its ability to survive and grow in hot water heaters. Further, the growth of *M. avium* at 42 °C leads to heightened survival at temperatures above 60 °C (Guenette et al., 2020). *M. avium* is also able to tolerate periods of stagnancy, common in premise plumbing, and can grow at 6% and 12% oxygen as well as air (21% oxygen) (Lewis and Falkinham, 2015). Although microbiological laboratories grow mycobacteria on rich media originally developed for the cultivation of *Mycobacterium tuberculosis*, the MAC can grow in water of low organic matter content. *M. avium* growth was measured in a pilot water distribution system containing 50 μg of assimilable organic carbon (AOC; Norton et al., 2005).

Another factor likely contributing to the residence of MAC in drinking water systems and premise plumbing is its resistance to killing by phagocytic amoebae; MAC is included in the group of amoeba-resisting microorganisms (ARMs) (Thomas et al., 2008). Rather than being phagocytized and killed by amoebae (*Acanthamoeba* spp., *Hartmanella* spp., and *Vermamoeba* spp.), ARMs multiply within amoebae (Thomas et al., 2008). The spectrum of characteristics of *M. avium* makes it ideally adapted to drinking water distribution systems and premise plumbing, such as disinfectant resistance, growth in drinking water, thermal tolerance, biofilm formation, growth under conditions of stagnation, and resistance to killing by amoebae.

The major determinant of many of the physiological features of members of the MAC is the thick, lipid-rich, hydrophobic outer membrane that makes up 30% of the weight of the mycobacterial cells (Brennan and Nikaido, 1995). The compilation of the advantages and disadvantages of the presence of that outer membrane is given in Table 4.2. The slow growth of MAC is due, in part, to the diversion of ATP to the synthesis of the C_{60}–C_{80} lipids in that outer membrane. It is important to point out that the slow increase in cell number of MAC strains is not due to slow metabolism, but to diversion of resources. MAC metabolism is as rapid as that of *Escherichia coli*; it simply diverts ATP to lipid synthesis, leading to cells that are staggeringly resistant to antimicrobial agents.

Table 4.2 Advantages and disadvantages of the lipid-rich outer membrane of MAC

Advantage	Disadvantage
Hydrophobic Cells	
Disinfectant Resistance	Reduced Uptake
Surface Adherence	Aggregation
Biofilm Formation	Slow Growth
Persistence in Flow	Adhere to Particles
Metal Resistance	Slow Metal Uptake
Aerosol Drop Concentration	
Desiccation Resistance	
Antibiotic Resistance	
Metabolize Organics	
Long Chain Lipid	
Hydrophobic Cells	Diversion of Energy
	Slow Growth

References

Adjemian J, Olivier KN, Seitz AE, Falkinham JO III, Holland SM, Prevots DR. (2012). Spatial clusters of nontuberculous mycobacterial lung disease in the United States. *Am J Respir Crit Care Med.* 186: 553–558.

Billinger ME, Olivier KN, Viboud C, Monets de Oca R, Steiner C, Holland SM, Prevots DR. 2009. Nontuberculous mycobacteria-associated lung disease in hospitalized persons, United States, 1998–2005. *Emerg Infect Dis.* 15: 1562–1569.

Brennan PJ, Nikaido H. (1995). The envelope of mycobacteria. *Annu Rev Biochem.* 64: 29–63.

Collier SA, Stockman LJ, Hicks LA, Garrison LE, Zhou FJ, Beach MJ. 2012. Direct healthcare costs of selected diseases primarily or partially transmitted by water. *Epidemiol Infect.* 140: 2003–2013.

De Groot MA, Pace NR, Fulton K, Falkinham JO III. 2006. Relationships between *Mycobacterium* isolates from patients with pulmonary mycobacterial infection and potting soils. *Appl Environ Microbiol.* 72: 7062–7606.

Drake TA, Herron RM, Hindler JA, Berlin OGW, Bruckner DA. 1988. DNA probe reactivity of *Mycobacterium avium* complex isolates from patients without AIDS. *Diagn Microbiol Infect Dis.* 11: 125–128.

Falkinham JO III. 2009. Surrounded by mycobacteria: Nontuberculous mycobacteria in the human environment. *J Appl Microbiol.* 107: 356–367.

Falkinham JO III. 2010. Hospital water filters as a source of *Mycobacterium avium* complex. *J Med Microbiol.* 59: 1198–1202.

Falkinham JO III. 2011. Nontuberculous mycobacteria from household plumbing of patients with nontuberculous mycobacterial disease. *Emerg Infect Dis.* 17: 419–424.

Falkinham JO III, Iseman MD, de Haas P, van Soolingen D. 2008. *Mycobacterium avium* in a shower linked to pulmonary disease. *J Water Health.* 6: 209–213.

Falkinham JO III, Norton CD, LeChevallier MW. 2001. Factors influencing numbers of *Mycobacterium avium, Mycobacterium intracellulare,* and other mycobacteria in drinking water distribution systems. *Appl Environ Microbiol.* 67: 1225–1231.

Falkinham JO III, Parker BC, Gruft H. 1980. Epidemiology of infection by nontuberculous mycobacteria. I. Geographic distribution in the eastern United States. *Am Rev Respir Dis.* 121: 931–937.

Falkinham JO III, Williams MD, Kwait R, Lande L. 2016. *Methylobacterium* spp. as an indicator for the presence or absence of *Mycobacterium* spp. *Intl. J. Mycobacteriol.* **5:** 240–243.

George KL, Parker BC, Gruft H, Falkinham JO III. 1980. Epidemiology of infection by nontuberculous mycobacteria. II. Growth and survival in natural waters. *Am Rev Respir Dis.* 122: 89–94.

Goring SM, Wilson JB, Risebrough NR, Gallagher J, Carroll S, Heap KJ, Obradovic M, Loebinger MR, Diel R. 2018. The cost of *Mycobacterium avium* complex lung disease in Canada, France, Germany, and the

United Kingdom: A nationally representative observational study. *BMC Health Services Res.* 18: 700. doi: 10.1186/s12913-018-3489-8.

Griffith DE, Aksamit T, Brown-Elliott BA, Catanzaro A, Daley C, Gordin F, Holland SM, Horsburgh R, Huitt G, Iademarco MF, Iseman M, Olivier K, Ruoss S, von Reyn CF, Wallace RJ Jr, Winthrop K. 2007. An official ATS/IDSA statement: Diagnosis, treatment, and prevention of nontuberculous mycobacterial diseases. *Am J Respir Crit Care Med.* 175: 367–416.

Guide SV, Holland SM. 2002. Host susceptibility factors in mycobacterial infection: Genetics and body morphotype. *Infect Dis Clin North Am.* 16: 163–186.

Guenette S, Williams MD, Falkinham JO III. 2020. Growth temperature, trehalose, and susceptibility to heat in *Mycobacterium avium*. *Pathogens.* 9: 657.

Haller S, Höller C, Jacobshagen A, Hamouda O, Abu Sin M, Monnet DL, Lachouras D, Eckmanns T. 2016. Contamination during production of heater-cooler units by *Mycobacterium chimaera* potential cause for invasive cardiovascular infections: Results of an outbreak investigation in Germany, April 2015 to February 2016. *Euro Surveill.* 21: 30215. doi.org/10.2807/1560-7917.ES.2016.21.17.30215.

Hamilton L, Falkinham JO III. 2018. Aerosolization of *Mycobacterium avium* and *Mycobacterium abscessus* from a household ultrasonic humidifier. *J. Med. Microbiol.* 67: 1491–1497. doi: 10.1099/jmm.0.000822.

Hatzenbuehler LA, Tobin-D'Angelo M, Drenzek C, Peralta G, Cranmer LC, Anderson EJ, Milla SS, Abramowicz S, Yi J, Hilinski J, Rajan R, Whitley MK, Gower V, Berkowitz F, Shapiro CA, Williams JK, Harmon P, Shane AL. 2017. Pediatric dental clinic–associated outbreak of *Mycobacterium abscessus* infection. *J Pediatric Infect Dis Soc.* 6: e116–e122.

Iakhiaeva E, McNulty S, Brown Elliott BA, Falkinham JO III, Williams MD, Vasireddy R, Wilson RW, Turenne C, Wallace RJ Jr. 2013. MIRU-VNTR of *Mycobacterium intracellulare* and *Mycobacterium chimaera* for strain comparison with establishment of a PCR-based data base. *J Clin Microbiol.* 51: 409–416.

Kahana LM, Kay JM, Yakrus MA, Waserman S. 1997. *Mycobacterium avium* complex infection in an immunocompetent young adult related to hot tub exposure. *Chest.* 111: 242–245.

Katsuda R, Yoshida S, Tsuyuguchi K, Kawamura T. 2018. A case report of hot tub lung: Identical strains of *Mycobacterium avium* from the patient and the bathroom air. *Int J Tubercul Lung Dis.* 22: 3520352.

Kim JS, Tanake N, Newell JD, De Groote MA, Fulton K, Huitt G, Lynch DA. 2005. Nontuberculous mycobacterial infection. CT scan findings, genotype, and treatment responsiveness. *Chest.* 128: 3363–3369.

Kirschner RA Jr, Parker BC, Falkinham JO III. 1992. Epidemiology of infection by nontuberculous mycobacteria. X. *Mycobacterium avium, M. intracellulare,* and *M. scrofulaceum* in acid, brown-water swamps of the southeastern United States and their association with environmental variables. *Am Rev Respir Dis.* 145: 271–275.

Kirschner RA Jr, Parker BC, Falkinham JO III. 1999. Humic and fulvic acids stimulate the growth of *Mycobacterium avium. FEMS Microbiol Ecol.* 30: 327–332.

Lande L, Alexander DC, Wallace RJ Jr, Kwait R, Iakhiaeva E, Williams M, Cameron ADS, Olshefsky S, Devon R, Vasireddy R, Peterson DD, Falkinham JO III. 2019. *Mycobacterium avium* in community and household water, suburban Philadelphia, Pennsylvania, USA, 2010–2012. *Emerg Infect Dis.* 25: 473–481.

Lewis AH, Falkinham JO III. 2015. Microaerobic growth and anaerobic survival of *Mycobacterium avium, Mycobacterium intracellulare,* and *Mycobacterium scrofulaceum. Int J Mycobacteriol.* 4: 25–30.

Maras TK, Chedore P, Ying AM, Jamieson F. 2007. Isolation prevalence of pulmonary non-tuberculous mycobacteria in Ontario, 1997–2003. *Thorax.* 62: 661–666.

Maras TK, Daley CL. 2002. Epidemiology of human pulmonary infection with nontuberculous mycobacteria. *Clin Chest Med.* 23: 553–567.

Mirsaeidi M, Allen MB, Ebrahimi G, Schraufnagel D. 2015. National hospital costs for pulmonary mycobacterial disease in the USA from 2001 to 2012. *Int J Mycobateriol.* 4: 156–157.

Mullis SN, Falkinham JO III. 2013. Adherence and biofilm formation of *Mycobacterium avium* and *Mycobacterium abscessus* to household plumbing materials. *J Appl Microbiol.* 115: 908–914.

Norton CD, LeChevallier MW, Falkinham JO III. 2004. Survival of *Mycobacterium avium* in a model distribution system. *Water Res.* 38: 1457–1466.

Parker BC, Ford MA, Gruft H, Falkinham JO III. 1983. Epidemiology of infection by nontuberculous mycobacteria. IV. Preferential aerosolization of *Mycobacterium intracellulare* from natural waters. *Am. Rev. Respir. Dis.* 128: 652–656.

Prince DS, Peterson DD, Steiner RM, Gottlieb JE, Scott R, Israel HL, Figueroa WG, Fish JE. 1989. Infection with *Mycobacterium avium* complex in patients without predisposing conditions. *New Engl J Med.* 321: 863–868.

Redelman-Sidi G, Sepkowitz KA. 2010. Rapidly growing mycobacteria infection in patients with cancer. *Clin Infect Dis.* 51: 422–434.

Sax H, Bloemberg G, Hasse B, Sommerstein R, Kohler P, Achermann Y, Rössle M, Falk V, Kuster SP, Böttger EC, Weber R. 2015. Prolonged outbreak of *Mycobacterium chimaera* infection after open-chest heart surgery. *Clin Infect Dis.* 61: 67–75. doi: 10.1093/cid/civ198.

Schelonka RL, Ascher DP, McMahon DP, Drehner DM. 1994. Catheter-related sepsis caused by *Mycobacterium avium* complex. *Ped Infect Dis J.* 13: 236–238.

Schulze-Röbbecke R, Buchholtz K. 1992. Heat susceptibility of aquatic mycobacteria. *Appl Environ Microbiol.* 58: 1869–1873.

Simpson GL, Raffin TA, Remington JS. 1982. Association of prior nocardiosis and subsequent occurrence of nontuberculous mycobacteriosis in a defined, immunosuppressed population. *J Infect Dis.* 146: 211–219.

Singh J, O'Donnell K, Ashouri N, Adler-Shohet FC, Nieves D, Tran MT, Arrieta A, Tran L, Cheung M, Zahn M. 2018. Outbreak of invasive nontuberculous *Mycobacterium* (NTM) infections associated with a pediatric dental practice. *Open Forum Infect Dis.* 5(Suppl 1): 926. doi: 10.1093/ofid/ofy209.067.

Steed KA, Falkinham JO III. 2006. Effect of growth in biofilms on chlorine susceptibility of *Mycobacterium avium* and *Mycobacterium intracellulare*. *Appl Environ Microbiol.* 72: 4007–4010.

Taylor RM, Norton CD, LeChevallier MW, Falkinham JO III. 2000. Susceptibility of *Mycobacterium avium, Mycobacterium intracellulare,* and *Mycobacterium scrofulaceum* to chlorine, chloramine, chlorine dioxide, and ozone. *Appl Environ Microbiol.* 66: 1702–1705.

Thomas V, Loret J-F, Jousset M, Greub G. 2008. Biodiversity of amoebae and amoebae-resisting bacteria in a drinking water treatment plant. *Environ Microbiol.* 10: 2728–2745.

Thomson RM, Armstrong JG, Looke DF. 2007. Gastroesophageal reflux disease, acid suppression, and *Mycobacterium avium* complex pulmonary disease. *Chest.* 131: 1166–1172.

Tichenor WS, Thurlow J, McNulty S, Brown-Elliott BA, Wallace RJ Jr, Falkinham JO III. 2012. Nontuberculous mycobacteria in household plumbing as possible cause of chronic rhinosinusitis. *Emerg Inf Dis.* 18: 1612–1617.

Torriani FJ, Behling CA, McCutchan JA, Haubrich RH, Havlir DV. 1996. Disseminated *Mycobacterium avium* complex: Correlation between blood and tissue burden. *J Infect Dis.* 173: 942–949.

Tortoli E. 2003. Impact of genotypic studies on mycobacterial taxonomy: The new mycobacteria of the 1990s. *Clin Microbiol Revs.* 16: 319–354.

Tortoli E, Rindi L, Garcia MJ, Chiardonna P, Dei R, Garzelli C, Kroppenstedt RM, Lari N, Mattei R, Mariottini A, Mazzarelli G, Murcia MI, Nanetti A, Piccoli P, Scarpaco C. 2004. Proposal to elevate the genetic variant MAC-A, included in the *Mycobacterium avium* complex, to species rank as *Mycobacterium chimaera* sp. *Int J Syst Evol Microbiol.* 54: 1227–1285.

Tzou CL, Dirac MA, Becker AL, Beck NK, Weigel KM, Meschke JS, Cangelosi GA. 2020. Association between *Mycobacterium avium* complex pulmonary disease and mycobacteria in home water and soil. A case–control study. *Ann Am Thorac Soc.* 17: 57–62. doi: 10.1513/AnnalsATS.201812-915OC

von Reyn CF, Arbeit RD, Horsburgh CR, Ristola MA, Waddell RD, Tvaroha SM, Samore M, Hirschhorn LR, Lumio J, Lein AD, Grove MR, Tosteson ANA. 2002. Sources of disseminated *Mycobacterium avium* infection in AIDS. *J Infect.* 44: 166–170.

Wallace RJ Jr, Iakhiaeva E, Williams MD, Brown-Elliott BA, Vasireddy S, Vasireddy R, Lande L, Peterson DD, Sawicki J, Kwait R, Tichenor WS, Turenne C, Falkinham JO III. 2013. Absence of *Mycobacterium intracellulare* and the presence of *Mycobacterium chimaera* in household water and biofilm samples of patients in the U.S. with *Mycobacterium avium* complex respiratory disease. *J Clin Microbiol.* 51: 1747–1752.

Wolinsky E. 1995. Mycobacterial lymphadenitis in children: A prospective study of 105 nontuberculous cases with long-term follow-up. *Clin Infect Dis.* 20: 954–963.

4.4 Amoebae and Protozoa

4.4.1 Introduction

Amoebae and protozoa are two heterogeneous groups of eukaryotic microorganisms. Some are motile using flagella or cilia (protozoa) and some move by contact with surfaces and lack distinct cell morphologies (amoebae). Both amoebae and protozoa have different cell stages, including cyst forms that provide resistance to environmental stresses, particularly allowing long-term survival under nutrient-free or unfavorable conditions, such as desiccation (Delafont et al., 2018). Some amoebae and protozoa are free-living, some are obligate pathogens. The focus here will be the free-living, opportunistic pathogens.

The waterborne amoebae that cause human infection include *Acanthamoeba* spp., *Vermamoeba* spp., *Hartmanella* spp., and *Naegleria fowleri.* All are free-living, opportunistic premise plumbing pathogens, although the majority of infections are attributable to *Acanthamoeba* spp. in the United States and elsewhere (Schuster and Visvesvara, 2004). *Naegleria fowleri* is not an opportunist in that infected individuals do not have a predisposing risk factor, i.e., behavior. *Naegleria fowleri* is a free-living, thermotolerant protozoan and causes headline-grabbing, life-threatening meningoencephalitis. It is also described as "Brain-Eating Amoeba" and will be discussed in this chapter.

Historically, amoebae and protozoa were identified on the basis of cellular morphology, motility, and the presence of alternative forms, namely trophozoites, flagellated or ciliated forms, and cysts (Delafont et al., 2018). However, that approach is time-consuming and requires skilled microbiologists to distinguish the different species. It has been shown that 18S rRNA sequences allow taxonomic groups of amoeba to be distinguished from one another (Lee et al., 2010; Chen et al., 2016; Sánchez et al., 2018). Because of that new approach, earlier reports of amoeba may carry names that have changed. For example, the differences of the 18S rRNA sequence of *Hartmanella vermiformis* led to its being renamed as *Vermamoeba vermiformis* (Delafont et al., 2018; Scheid, 2019).

For the purposes of this volume, I propose to ignore the protozoa, both the free-living and obligate parasites. Most of the pathogenic protozoa (e.g., *Giardia, Entamoeba, Toxoplasma, Cyclospora,* and *Cryptosporidium*) enter drinking water from other sources (e.g., human feces) and thereby can be transmitted by water. Thus, just like *E. coli* and *Salmonella*, they are contaminants, not colonists, of drinking water. These protozoa do not meet the criteria for inclusion as opportunistic premise plumbing pathogens and will not be discussed further. There is one free-living protozoan *Tetrahymena pyriformis* that is not pathogenic but has been used as a model eukaryotic and phagocytic microbe for studying phagocytosis and the survival, growth, and persistence of intracellular bacteria (Strahl et al., 2001). As the case for amoebae, taxonomy and identification of protozoa has historically involved morphological features, but is changing due to the advent of means to identify via 18S rRNA sequences (Chen et al., 2016; Lee et al., 2010; Sánchez et al., 2018).

4.4.2 Diseases and Risk Factors of Amoebae

Diseases caused by *Acanthamoeba* spp. and other amoebae include ocular keratitis in immunocompetent individuals and disseminated infections in the immunosuppressed, such as a consequence of HIV infection (Schuster and Visvesvara, 2004). The majority of infections are caused by *Acanthamoeba* spp. Ocular keratitis is quite commonly associated with the presence of *Acanthamoeba* spp. in ophthalmic solutions (Schuster and Visvesvara, 2004). By far, the most serious consequence of *Acanthamoeba* spp. is haematogenic spread throughout the body, particularly in immunodeficient individuals. In those instances, transmission in the body can lead to encephalitis and nasopharyngeal, cutaneous, and disseminated (bacteremia) infections (Schulster and Visvesvara, 2004). A useful adjunct to studies of free-living amoebae and *Acanthamoeba* spp. is a PCR-based method to distinguish pathogenic and nonpathogenic *Acanthamoeba* spp. (Howe et al., 1997).

4.4.3 Amoebae Ecology and Sources

Amoebae, including *Acanthamoeba, Hartmanella,* and *Vermamoeba,* surround humans. They have been isolated from soils, fresh and brackish waters, and sewage. Further, they are present in source water for treatment, distribution systems, and premise plumbing (Schuster and Visvesvara, 2004). Residence in drinking water leads to *Acanthamoeba* spp. presence in hospitals, hydrotherapy baths, and shower heads (Schuster and Visvesvara, 2004). Their presence in the human environment is reflected by the high frequency (87%) of antibodies to *Acanthamoeba* in humans (Schuster and Visvesvara, 2004). Free-living amoebae and *Acanthamoeba* spp. are present in source water, water treatment plants, and drinking water. In a study in Germany, free-living amoebae were found in raw water and in the treatment plant after flocculation, sedimentation, and filtration, and in groundwater (Hoffman and Michel, 2001). Further, in the United States, 13% of drinking water samples collected in Huntington, West Virginia had free-living amoebae and 9.3% had *Acanthamoeba* spp. in spite of 1.56 mg/L concentration of chlorine (Trzyna et al., 2010). Household water and biofilm samples yield high numbers of free-living amoebae and *Acanthamoeba* spp. as shown by an exceptional CDC study (Stockman et al., 2011). In Ohio, amoebae were found in 79% (371/467) households, with 52% of showerhead water samples and 50% of kitchen sprayers yielding amoebae (Stockman et al., 2011). High percentages of free-living amoebae (79%) and *Acanthamoeba* spp. (51%) were found in 2,454 water samples collected from homes (Stokman et al., 2011). Detection of amoebae was higher in swab samples of biofilms than water samples (Stockman et al., 2011). *Vermamoeba* (nee, *Hartmanella*), *Acanthamoeba,* and *Vahlkampfia* spp. were the most common isolated amoebae (Stockman et al., 2011). PCR-based methods have been described for the detection and enumeration of *Vermamoeba vermiformis* spp. (nee *Hartmanella* spp.; Kuiper et al., 2006) and *Acanthamoeba* spp. (Rivière et al., 2006).

4.4.4 Amoebae Transmission Pathways

4.4.4.1 Airway infection

Acanthamoeba spp. infection can occur by forced introduction of *Acanthamoeba*-containing water in the nasal cavity as a result of jumping into the water. Further, as children may often play in the water for long periods of time and can inhale or ingest water; this behavior is a risk factor for infection (Schulster and Visvesvara, 2004).

4.4.4.2 Dermal infection

As is the case for other opportunistic premise plumbing pathogens, dermal *Acanthamoeba* spp. infections can be a consequence of exposure of cuts to nature water (lakes, streams), potable water, and solutions that have been contaminated by *Acanthamoeba*.

4.4.4.3 Ocular infection

Ocular keratitis is quite commonly associated with the presence of *Acanthamoeba* spp. in ophthalmic solutions or poor hygiene in the care of soft contact lenses and the lens' cases (Schuster and Visvesvara, 2004). Often, ocular keratitis is associated with the use of non-sterile water to prepare saline solutions for lens care. Further, a case of keratitis was linked to an outdoor hot tub, demonstrating the role of the widespread distribution of *Acanthamoeba* spp. in water and their high frequency of thermotolerance (Samples et al., 1984).

4.4.5 Physiologic Ecology

A study of Paris tap water samples showed that free-living amoebae were particularly high at distribution system endpoints (67%) compared to reservoir-sampling sites (Delafont et al., 2016). *Vermamoeba* spp. (nee *Hartmanella* spp.), *Acanthamoeba* spp., and *Echinamoeba* spp. were the most prevalent (88%) of the free-living amoebae detected (Delafont et al., 2016). Water source, total chlorine concentration, pH, temperature, and conductivity all influenced the numbers of free-living amoebae (Delafont et al., 2016). Specifically, *Acanthamoeba* spp. presence was correlated with high pH, *Vermamoeba* spp. with high temperature, and *Echinamoeba* spp. with higher conductivity (Delafont et al., 2016). Water sources also influenced free-living amoebae; for example, *Echinamoeba* spp. was more prevalent in samples from underground water (Delafont et al., 2016). Two bacterial genera, *Pseudomonas* spp. and *Stenotrophomonas* spp were detected more frequently in association with free-living amoebae compared to other bacteria (Delafont et al., 2016).

A major feature of *Acanthamoeba* spp. is the formation of thick-walled cysts, which are resistant to disinfectants, desiccation, heat, and ultraviolet light (Schulster and Visvesvara, 2004). The cysts are more resistant to chlorine, tolerating 50 mg/L, than the vegetative growing form called trophozoites, which can tolerate only 4 chlorine mg/L (Trzyna et al., 2010). Cyst formation ensures long-term survival under stressful conditions.

A note has been made of the isolation of thermotolerant strains of *Acanthamoeba* spp. (Schuster and Visvesvara, 2004). Mostly all (90%) of isolates of *Acanthamoeba* spp. recovered from ponds in Oklahoma (USA) in summer proved to be thermophilic (John and Howard, 1996). A similar result was obtained of *Acanthamoeba* spp. isolates from river and reservoir water in Germany (Hoffman and Michel, 2001).

4.4.6 *Naegleria fowleri*

4.4.6.1 *Naegleria fowleri* diseases and risk factors

N. fowleri is a free-living amoeba that causes a particular infection, meningoencephalitis, which is almost uniformly fatal. The normal habitats of *N. fowleri* include natural soils and waters and infection is associated with jumping in ponds and lakes (Yoder et al., 2010). Infections are most frequent in young boys (median age 10–14 years) during the summer (July–September) (Yoder et al., 2010). A case of primary amebic meningoencephalitis due to *N. fowleri* was reported in St. Thomas, the U.S. Virgin Islands in 2012 (MMWR, 2013). The patient was 47 years old and had practiced nasal rinsing, presented at the emergency department with a headache, but the following day presented with fever, confusion, agitation, and a severe headache (MMWR, 2013). Five days after the initial visit to the emergency department, the patient died (MMRS, 2013). Three water samples (i.e., water heater, shower water, and showerhead) of 17 water samples collected from the patient's home yielded *N. fowleri* (MMWR, 2013). The home was fed with rainwater and untreated groundwater from a well (MMWR, 2013). To date, no genetic factors have been associated with a higher risk of infection; the major risk appears to be jumping into that water.

Most of the reported cases have occurred in the southern United States, from California to Virginia with the highest number in Texas and Florida (Yoder et al., 2010). The route of infection is through the nasal cavity and it is presumed that jumping into water forces water up into the nasal cavity and brain (Yoder et al., 2010). The study of *N. fowleri* epidemiology is difficult because of the rarity of infections. Over the period 1962–2008, only 111 cases have been diagnosed (Yoder et al., 2010).

4.4.6.2 *Naegleria fowleri* ecology, sources, and transmission

The source for human infection by *N. fowleri* includes natural waters (Yoder et al., 2010) and untreated well water (MMWR, 2013). However, a 2020 report documented *N. fowleri* infection and death in 6-year old boy that was traced to the presence of *N. fowleri* in the water of the "splash-pad" in Texas City where he lived. It shows that *N. fowleri* is not restricted to natural or untreated well waters, but must also survive and persist in treated drinking water.

N. fowleri feeds on bacteria and can be attacked by other free-living amoebae. Further, it grows at 42 °C (Červa, 1978), which is consistent with its preference for the water of high temperature, and does not grow in growth medium containing 0.1 N NaCl (0.585%) (Červa, 1977), so is unlikely to reside in ocean water or estuaries. In a freshwater biofilm culture at a temperature of 42 °C, *N. fowleri* reached densities of 30–900 cells/cm^2 (Goudot et al., 2012). However, growth appeared to be within a narrow temperature range, as very low numbers of *N. fowleri* were found in the same biofilms incubated at 32 °C (Goudot et al., 2012). That behavior suggests *N. fowleri* is a thermophile.

References

Carnt NA, Subedi D, Connor S, Kilvington S. 2020. The relationship between environmental sources and the susceptibility of *Acanathamoeba* keratitis in the United Kingdom. *PLoS ONE*. 15: 30229681. doi: https://doiorg/10.1371/journal.pone.0229681.

Červa L. 1978. Some further characteristics of the growth of *Naegleria fowleri* and *N. gruberi* in axenic culture. *Folia Parasit.* 25: 1–8.

Červa L. 1977. The influence of temperature on the growth of *Naegleria fowleri* and *N. gruberi* in axenic culture. *Folia Parasit.* 24: 221–228.

Chen M-X, Ai L, Chen J-H, Feng X-Y, Chen S-H, Cai Y-C, Lu Y, Zhou X-N, Chen J-X, Hu W. 2016. DNA microarray detection of 18 important human blood protozoan species. *PLoS Negl Trop Dis.* 10: e0005160.

Delafont V, Rodier M-L-H, Maisonneuve E, Cateau E. 2018. *Vermamoeba vermiformis*: A free-living amoeba of interest. *Microb Ecol.* 76: 991–1001.

Goudot S, Herbelin P, Mathieu L, Soreau S, Banas S, Jorand F. 2012. Growth dynamic of *Naegleria fowleri* in a microbial freshwater biofilm. *Water Res.* 46: 3958–3966.

Hoffman R, Michel R. 2001. Distribution of free-living amoebae (FLA) during preparation and supply of drinking water. *Int J Hyg Environ Hlth.* 203: 215–219.

Howe DK, Vodkin MH, Novak RJ, Visvesvara G, McLaughlin GL. 1997. Identification of two genetic markers that distinguish pathogenic and nonpathogenic strains of *Acanthamoeba* spp. *Parasitol Res.* 83: 345–348.

John DT, Howard MJ. 1996. Techniques for isolating thermotolerant and pathogenic free-living amoebae. *Folia Protistol.* 43: 267–271.

Kuiper MW, Valster RM, Wullings BA, Boonstra H, Smidt H, van der Kooij D. 2006. Quantitative detection of the free-living amoeba *Hartmanella vermiformis* in surface water using real-time PCR. *Appl Environ Microbiol.* 72: 5750–5756.

Lee D-Y, Seto P, Korczak R. 2010. DNA microarray-based detection and identification of waterborne protozoan pathogens. *J Microbiol Meth.* 80: 129–133.

MMWR, Morbidity and Mortality Weekly Report. 2013. Notes from the field: Primary amebic meningoencepthalitis associated with ritual nasal rinsing—St. Thomas, U.S. Virginia Islands, 2012. 62: 903–903.

Rayamajhee B, Subedi D, Won S, Kim JY, Vijay A, Tan J, Henriquez FLl, Willcox M. Carnt NA. 2020. Investigating domestic shower settings as a risk factor for *Acanthamoeba* keratitis. *Water.* 12: 3493. doi:10.3390/w12123493.

Rivière D, Szczebara FM, Berjeaud J-M, Frère J, Héchard Y. 2006. Development of a real-time PCR assay for quantification of *Acanthamoeba* trophozoites and cysts. *J Microbiol Methods.* 64: 78–83.

Samples JR, Binder PS. Luibel FJ, Font RL, Visvesvara GS, Peter CR. 1984. *Acanthamoeba* keratitis possibly acquired for a hot tub. *Arch Ophthalmol.* 102: 707–710.

Sánchez C, López MC, Galeano LA, Qvarnstrom Y, Houghton K, Ramírez JD. 2018. Molecular detection and genotyping of pathogenic protozoan parasites in raw and treated water samples from southwest Columbia. *Parasites Vectors.* 11: 563.

Scheid PL. 2019. *Vermamoeba vermiformis*: A free-living amoeba with public health and environmental health significance. *Open Patrasitol J.* 7: 40–47.

Schuster FL, Visvesvara GS. 2004. Free-living amoebae as opportunistic and non-opportunistic pathogens of humans and animals. *Int J Parasitol.* 34: 1001–1027.

Stahl LM, Olson JB. 2020. Environmental abiotic and biotic factors affecting the distribution and abundance of *Naegleria fowleri*. *FEMS Microb Ecol.*

Strahl ED, Gillaspy GE, Falkinham JO III. 2001. Fluorescent acid-fast microscopy for measuring phagocytosis of *Mycobacterium avium, Mycobacterium intracellulare*, and *Mycobacterium scrofulaceum* by *Tetrahymena pyriformis* and their intracellular growth. *Appl. Environ. Microbiol.* 67: 4432–4439.

Stockman LJ, Wright CJ, Visvesvara GS, Fields BS, Beach MJ. 2011. Prevalence of *Acanthamoeba* spp. and other free-living amoebae in household water, Ohio, USA—1990–1992. *Parasitol Res.* 108: 621–627.

Trzyna WC, Mbugua MW, Rogerson A. 2010. *Acanthamoeba* in the domestic water supply of Huntington, West Virginia, U.S.A. *Acta Protozool.* 49: 9–15.

Yoder JS, Eddy BA, Wisvesvara GS, Capewell L, Beach MJ. 2010. The epidemiology of primary amoebic meningoencephalitis in the USA, 1962–2008. *Epidemiol Infect.* 138: 968–975.

Chapter 5

Opportunistic Premise Plumbing Pathogens as Amoebae-Resisting Microorganisms

5.1 Introduction to Amoebae-Resisting Bacteria (ARB)

Amoebae-resisting microorganisms (ARMs) or protozoa-resisting bacteria (PRBs) are bacteria that benefit from their interaction with free-living amoebae, protozoa, or nanoflagellates (Thomas et al., 2010). Both groups share the same habitat, namely natural water and soils and water in engineered systems. The benefits for ARMs include (a) survival in spite of phagocytosis, (b) intracellular growth, (c) survival in spite of encystment by hosts, and in some instances (d) increases in virulence (Thomas et al., 2010).

5.1.1 Phagocytic Microorganisms

Although a great deal of attention has been placed on the phagocytic amoebae as harboring opportunistic premise plumbing pathogens (OPPPs), other groups of waterborne microorganisms, as illustrated below, carry OPPPs. These include protozoa, such as the waterborne flagellated *Tetrahymena pyriformis*, and the

Opportunistic Premise Plumbing Pathogens
Joseph O. Falkinham, III

ISBN 978-981-4968-40-9 (Hardcover), 978-1-003-32100-2 (eBook)
www.jennystanford.com

predatory nanoflagellates; for example, the marine *Paraphysomonas* spp., the marine *Ochromonas* spp., and the freshwater *Spumella* spp.

5.1.2 ARB–PRB

In Table 5.1, a list of OPPPs that have been reported as ARMs is given.

Table 5.1 OPPPs and ARMs

Legionella pneumophila
Mycobacterium avium complex (MAC)
Pseudomonas aeruginosa
Stenotrophomonas maltophilia
Acinetobacter baumannii
Sphingomonas paucimobilis
Methylobacterium extorquens

5.2 Shared Habitats of Amoebae and OPPPs

5.2.1 Coincidence of *L. pneumophila* and Amoebae

A wide range of drinking water habitats used by humans contain amoebae and OPPPs. A review of the presence of free-living amoebae in treated drinking water reported that approximately 50% of samples from treated water in samples from European distribution systems had free-living amoebae (Thomas and Ashbolt, 2011). Those free-living amoebae were also detected at the point-of-use of the treated water (Thomas and Ashbolt, 2011). Hospital water samples from a hospital in Lausanne, Switzerland, yielded both amoebae [e.g., *Vermamoeba vermiformis* (nee, *Hartmanella vermiformis*) and *Acanthamoeba polyphaga*] and OPPPs (e.g., *Pseudomonas aeruginosa, Legionella pneumophila, Mycobacterium xenopi*, and *Mycobacterium kansasii*) (Thomas et al., 2006). Water samples collected from the Seine River and a treatment plant 30 km upstream of Paris, France, and its distribution system and premise plumbing yielded both amoebae (e.g., *Acanthamoeba* spp., *Echinamoeba* spp., *Vannella* spp., and *Harmanella* spp.) and OPPPs (e.g., *Mycobacterium* spp., *Legionella* spp., *Pseudomonas* spp., and *Acinetobacter* spp.) (Thomas et al., 2008). Of 50 free-living amoebae isolates, 4 (8%) yielded bacteria, including *L. pneumophila* (Thomas et al., 2008).

A recent publication has reinforced the role of free-living amoebae and specifically, *Mycobacterium* spp., in the Paris drinking water system. A very high percentage of amoebal cultures (87.6%) carried *Mycobacterium* spp. (Delafont et al., 2014). Isolates of *Acanthamoeba, Vermamoeba, Echinamoeba,* and *Protoacanthamoeba* were found to carry *Mycobacterium* spp. showing that amoebal carriage of *Mycobacterium* spp. was not limited to one amoebal species (Delafont et al., 2014).

Free-living amoebae may also be responsible for the presence of *L. pneumophila* in dental unit waterlines. *Legionella*-laden amoebae were detected by cultivation and microscopy in 20% of dental unit waterlines (Singh and Coogan, 2005).

5.2.2 Growth of *Legionella pneumonia* in Amoebae

Based on the difficulty in reproducibly cultivating *L. pneumonia*, in spite of clear evidence of an infection source and the co-occurrence of *L. pneumophila* and amoebae, free-living amoebae are considered as hosts. Although amoebae are capable of utilizing bacteria and legionella as nutrient, it was shown that *L. pneumophila* could survive and grow in *A. castellanii* (Holden et al., 1984). Growth of *L. pneumophila* did not occur in amoebae-free medium or when amoebae and legionella were separated by a 0.4 μm pore size membrane demonstrating that growth was not reliant upon extracellular growth factors (Holden et al., 1984). Although debatable, it may be that *L. pneumophila* growth and survival in drinking water may be totally dependent upon intracellular growth in amoebae and other phagocytic, free-living microorganisms. The carriage of viable *L. pneumophila* by amoebae can be quite long, up to 850 days in one study (Shaheen et al., 2019) and protects the legionella from disinfection (Thomas et al., 2004).

5.3 OPPPs Isolated by Amoebal Co-Culture

5.3.1 OPPP Co-Culture

Based on the coincidence of amoebae and OPPPs, a number of investigators have used amoebae to isolate OPPPs. The theory behind those experiments has been to use the amoebae as efficient scavengers of bacterial prey. Co-cultivation of a sputum sample with *Acanthamoeba polyphaga* led to the recovery of *Legionella anisa*, an isolate that had not been isolated by conventional direct culture (La Scola et al., 2001). A high proportion of amoebae co-cultures of source water samples (42%), shower swabs (52%), and tap water swabs (45%) yielded OPPPs (Thomas et al., 2008). In that study, a lysate of an indigenous free-living *Vermamoeba vermiformis* (nee, *Hartmanella vermiformis*) strain yielded isolates of *Legionella pneumophila* (Thomas et al., 2006). It is important to point out in that case that the strain of *Vermamoeba vermiformis* was an isolate from a natural water sample, and not from a laboratory culture collection that could have been contaminated by *L. pneumophila*.

5.3.2 Resuscitation of Viable but Nonculturable (VBNC) *L. pneumophila* by *A. castellanii*

In addition to serving as a host for the replication and increased survival of *L. pneumophila, A. castellanii* is able to resuscitate its VBNC cells (Steinert et al., 1997). Although cells of *A. castellanii* were able to resuscitate *L. pneumophila*, based on the recovery of colony-forming units, the passage of the same VBNC *L. pneumophila* cells by intraperitoneal injection of guinea pigs did not result in resuscitation (Steinert et al., 1997).

5.3.3 Residence of *Mycobacterium* spp. in Protozoa

The survival and growth of OPPPs are not limited to amoebae. Cells of *Tetrahymena pyriformis* readily phagocytize the cells of *M. avium, M. intracellulare*, and *M. scrofulaceum*, and the intracellular *Mycobacterium* spp. cells grow by factors of 4–40-fold by 5 days in *T. pyriformis* cells (Strahl et al., 2001). The lowest increase of growth (4-fold) was shown by *M. scrofulaceum*, a weakly virulent *Mycobacterium* spp. compared to *M. avium* and *M. intracellulare* (Strahl et al., 2001). As these mycobacteria are intracellular pathogens, we have hypothesized that growth in protozoa and, by extension amoebae, may reflect differences in human virulence.

5.4 Physiologic Ecology of Amoebae–OPPPs Interaction

5.4.1 Predation-Pressure and Nutrient Availability Influence PRB

Although both *Pseudomonas aeruginosa* and *Mycobacterium* spp. are protozoa-resisting microorganisms, their numbers respond to predation-pressure and nutrient availability in different directions. *Mycobacterium* spp. numbers increased under protozoan predation pressure, while *P. aeruginosa* numbers fell (Andersson et al., 2018). In contrast under conditions of high nutrient availability, *P. aeruginosa* numbers increased and *Mycobacterium* spp. numbers fell (Andersson et al., 2018). This is consistent with observations suggesting that the environmental mycobacteria are oligotrophs, growing well at low-nutrient concentrations.

5.4.2 Determinants of Phagocytosis

The residence of OPPPs in free-living amoebae protozoa and predatory nanoflagellates requires phagocytosis of the bacteria and survival and intracellular growth. Two groups have investigated the effects of hydrophobicity and surface charge on the feeding rates of nanoflagellates. There was no influence of surface charge on the feeding rates of bacteria by the nanoflagellates (Matz and Jürgens, 2001), but the reports differed in the role of hydrophobicity, which is a factor in the phagocytosis of bacteria by macrophages (van Oss et al., 1975).

In one report, the highest bacterial grazing rates were found for cells with high hydrophobicity values (Monger et al., 1999). In contrast, the hydrophobicity of a mixed population of cells did not fall after a period of nanoflagellate grazing (Matz and Jürgens, 2001). The difference in the influence of hydrophobicity on grazing could have been due to the differences in methods and microbial populations.

5.4.3 Coexistence of Amoebae and *L. pneumophila*

Evidence from studies of *L. pneumophila* and free-living amoebae demonstrated that *L. pneumophila* was rare in copper tubing (Lu et al., 2014). However, persistence under conditions of low chlorine and low flow was associated with high numbers of free-living protozoa and amoebae (Lu et al., 2014). They speculated that as copper-piping is prevalent in many buildings and homes, amoebae and protozoa might serve to maintain *L. pneumophila* numbers (Lu et al., 2014).

5.5 Consequences of Residence of OPPPs in Amoebae

5.5.1 Growth Stimulation of *T. pyriformis* by *M. avium*

A unique characteristic of some strains of *T. pyriformis* is that it cannot grow in nutrient medium from the low-cell densities (i.e., 250 cells/mL or less (Schousbof and Rasmussen, 1994). However, they can grow to produce dense cultures if the medium is supplemented with a variety of compounds including fatty acids (Schousbof and Rasmussen, 1994). Thinking that mycobacterial cells are rich in fatty acids, the growth from the low-cell density of *T. pyriformis* cells carrying intracellular *M. avium* (Strahl et al., 2001) were diluted and used as an inoculum and their growth followed. *T. pyriformis* cells carrying *M. avium* were able to produce dense turbid cultures from low-cell densities (i.e., 1–100 cells/mL) compared to *T. pyriformis* cells not carrying *M. avium* (Falkinham, unpublished). These experimental results suggest a selective advantage to phagocytosis by protozoa, growth stimulation at low-protozoal densities.

5.5.2 Effect on OPPPs Carried by Amoebae

5.5.2.1 Increased ARM survival upon intra-amoebal residence

A wide variety of *Mycobacterium* species, including rapidly and slowly growing and pathogenic or saprophytic species, survived at least five days in *Acanthamoeba polyphaga* trophozoites and at least 15 days in cysts (Adékambi et al., 2006). Further, *Mycobacterium* spp. survived 24 h exposure to 15 gm chlorine/L (Adékambi et al., 2006). Thus, the intracellular residence of *Mycobacterium* spp. and by extension other OPPPs permit survival following chlorine exposure and long-term survival in spite of environmental stresses. This data was confirmed by the observation that the monochloramine-inactivation kinetics of intracellular *M. avium* reflected those of *A. castellanii*, and not those of *M. avium* (Berry et al., 2010).

5.5.2.2 Increased ARM virulence upon intra-amoebal residence

M. avium is phaogcytized by *Acanthamoeba castellanii* and replicates (Cirillo et al., 1997). Further, the *A. castellanii*-grown *M. avium* cells were taken up by *A. castellanii* more readily (8–10-fold) and were more virulent in the beige mouse model of infection (Cirillo et al., 1997).

5.5.2.3 Growth of *L. pneumophila* in amoebae

Different strains of *L. pneumophila* show different responses to growth temperature in association with amoebae (Buse and Ashbolt, 2011). Using A. polyphaga as host some *L. pneumophila* strains grew at 30 °C, 32 °C, and 37 °C, namely Bloomington-2, whereas others did not, namely Philadelphia-1 (Buse and Ashbolt, 2011). However, in *H. vermiformis*, Bloomington-2 failed to grow, suggesting that the choice of amoebae for co-cultivation or resuscitation will be *L. pneumophil*a strain-specific.

References

Adékambi T, Ben Salah S, Khlif M, Raoult D, Drancourt M. 2006. Survival of environmental mycobacteria in *Acanthamoeba polyphaga*. *Appl Environ Microbiol.* 72: 5974–5981.

Andersson A, Ahlinder J, Mathisen P, Hägglund, Bäckman S, Nilsson E, Sjökin A, Thelaus J. 2018. Predators and nutrient availability favor protozoa-resisting bacteria in aquatic systems. *Sci Rpts.* 8: 8415.

Berry D, Horn M, Xi C, Raskin L. 2010. *Mycobacterium avium* infections of *Acanthamoeba* strains: Host strain variability, grazing-acquired infections, and altered dynamics of inactivation with monochloramine. *Appl Environ Microbiol.* 76: 6685–6688.

Buse HY, Ashbolt NJ. 2011. Differential growth of *Legionella pneumophila* strains within a range of amoebae at various temperatures associated with in-premise plumbing. *Lttrs Appl Microbiol.* 53: 217–224.

Cirillo JD, Falkow S, Tompkins LS, Bermudez LE. 1997. Interaction of Mycobacterium avium with environmental amoebae enhances virulence. *Infect Immun.* 65: 3759–3767.

Delafont V, Mougari F, Cambau E, Joyeux M, Bouchon D, Héchard Y, Moulin L. 2014. First evidence of amoebae-mycobacteria association in drinking water network. *Environ Sci Technol.* 48: 11872–11882.

Holden EP, Winkler HH, Wood DO, Leinbach ED. 1984. Intracellular growth of *Legionella pneumophila* within *Acanthamoeba castellanii* Neff. *Infect Immun.* 45: 18–24.

La Scola B, Mezi L, Weiller PJ, Raoult D. 2001. Isolation of *Legionella anisa* using an amoebic coculture procedure. *J Clin Microbiol.* 39: 365–366.

Lu J, Buse HY, Gomez-Alvarez V, Struewing I, Santo Domingo J, Ashbolt NJ. 2014. Impact of drinking water conditions and copper materials on downstream biofilm microbial communities and *Legionella pneumophila* colonization. *J Appl Microbiol.* 117: 905–918.

Matz C, Jürgens K. 2011. Effects of hydrophobic and electrostatic cell surface properties of bacteria on feeding rates of heterotrophic nanoflagellates. *Appl Environ Microbiol.* 67: 814–820.

Monger BC, Landry MR, Brown SL. 1999. Feeding selection of heterotrophic marine nanoflagellates based on the surface hydrophobicity of their picoplankton prey. *Limnol Oceanogr.* 44: 1917–1927.

Schousboe P, Rasmussen L. 1994. Survival of *Tetrahymena thermophila* at low initial cell densities: Effects of lipids and long-chain alcohols. *J Euk Microbiol.* 41: 195–199.

Shaheen M, Scott C, Ashbolt NJ. 2019. Long-term persistence of infectious *Legionella* with free-living amoebae in drinking water biofilms. *Int J Hyg Environ Hlth.* 222: 678–686.

Singh T, Coogan MM. 2005. Isolation of pathogenic *Legionella* species and legionella-laden amoebae in dental unit waterlines. *J Hosp Infect.* 61: 257–262.

Steinert M, Emödy L, Amann R, Hacker J. 1997. Resuscitation of viable but unculturable *Legionella pneumophila* Philadelphia JR32 by *Acanthamoeba castellanii. Appl Environ Microbiol.* 63: 2047–2053.

Strahl ED, Gillaspy GE, Falkinham JO III. 2001. Fluorescent acid-fast microscopy for measuring phagocytosis of *Mycobacterium avium, Mycobacterium intracellulare,* and *Mycobacterium scrofulaceum* by *Tetrahymena pyriformis* and their intracellular growth. *Appl Environ Microbiol.* 67: 4432–4439.

Thomas JM, Ashbolt NJ. 2011. Do free-living amoebae in treated drinking water systems present an emerging health risk? *Environ Sci Technol.* 45: 860–869.

Thomas V, Bouchez T, Nicolas V, Robert S, Loret JF, Lévi Y. 2004. Amoebae in domestic water systems: Resistance to disinfection treatments and implication in *Legionella* persistence. *J Appl Microbiol.* 97: 950–963.

Thomas V, Herrera-Rimann K, Blanc DS, Greub G. 2006. Biodiversity of amoebae and amoebae-resisting bacteria in a hospital water network. *Appl Environ Microbiol.* 72: 2428–2438.

Thomas V, Loret J-F, Jousset M, Greub, G. 2008. Biodiversity of amoebae and amoebae-resisting bacteria in a drinking water treatment plant. *Environ Microbiol.* 10: 2728–2745.

Thomas V, McDonnell G, Denyer SP, Maillard J-Y. 2010. Free-living amoebae and their intracellular pathogenic microorganisms: Risks for water quality. *FEMS Microbiol Rev.* 34: 231–259.

Chapter 6

Common Features and Sources of Opportunistic Premise Plumbing Pathogens

6.1 Introduction

The list of characteristics shared by opportunistic premise plumbing pathogens (OPPPs) (Table 6.1) and identified in Chapters 4 and 5 defines a group of opportunistic pathogens that simply love engineered water distribution systems, including, most importantly, premise plumbing. Premise plumbing delivers drinking water in single-detached homes, apartments and condominiums, office buildings, and, most importantly, hospitals and healthcare centers. Thus, humans are surrounded by OPPPs. The physiology of this family of microorganisms defines their ecology; what my colleague, Professor Bruce Parker and I call their physiological ecology. Physiological ecology includes those physiological features that dictate the habitats occupied by OPPPs. Those common features are listed in Table 6.1 lead to the selection for OPPPs in premise plumbing systems. In addition to being opportunistic pathogens that persist and grow in premise plumbing, the characteristics of OPPPs that contribute to that behavior include (a) disinfectant resistance, (b) surface adherence and biofilm formation, (c) growth under low-nutrient

Opportunistic Premise Plumbing Pathogens
Joseph O. Falkinham, III

ISBN 978-981-4968-40-9 (Hardcover), 978-1-003-32100-2 (eBook)
www.jennystanford.com

(oligotrophic) conditions, (d) growth at low-oxygen concentrations, (e) thermal tolerance, (f) desiccation tolerance, (g) growth in amoeba and protozoa, and (h) non-mutational adaptation to stressful conditions.

Table 6.1 Common characteristics of OPPPs

Disinfectant Resistance
Surface Adherence and Biofilm Formation
Growth under Low-Carbon (Oligotrophic) Conditions
Growth under Low-Oxygen Conditions (Stagnation)
Thermal Tolerance
Desiccation Tolerance
Growth in Amoebae as Reservoirs
Adaptation to Stressful Conditions

6.2 Shared OPPP Characteristics Relevant to Survival, Persistence, and Growth in Premise Plumbing

6.2.1 Disinfectant Resistance

Compared to *E. coli*, the product of the chlorine concentration and length of exposure, called the $CT_{99.9\%}$, for *Escherichia coli* is 0.05 (Taylor et al., 2000). The $CT_{99.9\%}$ is 1,050-fold higher for *L. pneumophila* (Kuchta et al., 1985), 2,000-fold higher for *M. avium* (Taylor et al., 2000), and 40-fold higher for *P. aeruginosa* (Grobe et al., 2001). Thus, standard disinfectant dosages for drinking water treatment will essentially kill off all but the OPPPs. Consequently, even the relatively slower-growing OPPPs will be able to consume the available organic carbon without competition.

6.2.2 Surface Adherence and Biofilm Formation

Biofilm formation provides a mechanism to prevent the washing out of microorganisms in flowing systems, such as water-distribution systems and premise plumbing pipes. Surface adherence and biofilm formation and have been documented for *L. pneumophila* (Schofield and Locci, 1985), *M. avium* (Steed and Falkinham, 2006), and *P. aeruginosa* (de Beer et al., 1994). Further, for all three, residence in biofilms substantially increases their resistance to disinfectants (Cooper and Hanlon, 2010; deBeer et al., 1994; Steed and Falkinham, 2007). Measurements of the surface density of adherent cells (colony-forming units/cm^2) of those three OPPPs are as high as 10,000–15,000 CFU/cm^2 for *M. avium* (Mullis and Falkinham, 2017), *L. pneumophila* (Schofield and Locci, 1985), and *P. aeruginosa* (DeBeer et al., 1994). The magnitude of those values compared to typical CFU/mL of water samples from premise plumbing that are between 100 and 1,000, proves that the preferred habitat of the OPPPs are surfaces. That informs us to sample surface biofilms of distribution systems, premise plumbing, and medical equipment, rather than sample bulk water. Simply, the probability of finding an OPPP is much higher in a biofilm sample. Adherence of OPPPs to surfaces is quite rapid, driven in part by hydrophobicity (van Loosdrecht et al. 1989). OPPP cells are quite hydrophobic and that drives their preference for surface attachment over water suspension where the entire cell surface must interact with highly charged water. As expected, one sign of cell surface hydrophobicity is the clumping of cells in laboratory cultures.

The speed of surface attachment is illustrated by the behavior of *Mycobacterium chimaera* (a member of the *M. avium* complex) upon inoculation into a heater-cooler reservoir to compare different disinfection protocols. Following inoculation and an immediate 5 min circulation of water, only 0.1% of the inoculum could be recovered from the water; 99.9% of the cells were lost from suspension because of adherence to the surfaces in the heater-cooler (Falkinham, 2020).

6.2.3 Growth in Premise Plumbing at Low Organic Carbon Concentrations

Quite possibly, low organic carbon concentrations in drinking water may not limit the growth of all OPPPs. For example, long-term survival of *L. pneumophila* was documented under low-nutrient conditions (Pasko-Kolva et al., 1992) and *M. avium* thrives at assimilable organic concentrations (AOC) as low as 50 μg/L (Norton et al., 2005). The growth of *M. avium* is stimulated by humic and fulvic acids that are present in most surface water (Kirchner et al., 1999; Norton et al., 2005). *P. aeruginosa* can even grow in distilled water (Favero et al., 1971). As it is likely, in part, that *L. pneumophila* persists in drinking water systems and premise plumbing in amoebae, rather than as free-living cells (Thomas et al., 2008), its growth in the water of low-organic concentrations is dependent upon growth by phagocytosis of other microorganisms by amoebae.

6.2.4 Growth at Low-Oxygen Concentrations

Periods of stagnation in premise plumbing can increase OPPP numbers, as OPPPs are capable of growth even at low-oxygen levels, though not under conditions of anaerobiosis. For example, *L. pneumophila* grew at low-oxygen concentrations (Mauchline et al., 1992) and *L. pneumophila* numbers did not fall in hot water heaters when not in use and water was stagnating (Ciesielski et al., 1984), *M. avium* can grow at 6% and 12% oxygen (Lewis and Falkinham, 2015) and *P. aeruginosa* can grow in the absence of oxygen using nitrate as a terminal electron acceptor (Palmer et al., 2007). It might prove to be the case that high-oxygen concentrations would lead to the reduction in numbers of these microaerobic OPPPs. High-oxygen concentrations were associated with low *Mycobacterium* spp. numbers in surface water (Brooks et al., 1984; Kirchner et al., 1992), and the same holds true for *L. pneumophila* (Ciesielski et al., 1984).

6.2.5 High-Temperature Tolerance

Compared to contaminants of drinking water, namely *Escherichia coli* and other fecal-origin bacteria, the OPPPs are more tolerant of high temperatures. The temperature range of interest is that for the heated water. Generally, water heater temperatures are set between 120 °F (45 °C) and 130 °F (55 °C).

OPPPs grow in water heaters and can be found at layers where the temperature is optimal for survival and growth. That is particularly the case for electric water heaters with their distinctive temperature strata. The role of water heater temperature in maintaining OPPP numbers in premise plumbing was first discovered *Legionella* (Straus et al., 1996). Based on that report, in a study of *Mycobacterium* spp. in premise plumbing of *Mycobacterium* spp.-infected individuals, it was found that homes with a water heater temperature of greater than or equal to 125 °F (50 °C) were more likely to have *Mycobacterium* spp. in water samples than homes with a water heater temperatures equal to or higher than 130 °F (55 °C) (Falkinham, 2011). That information led to the recommendation that patients infected or at risk of *Mycobacterium* spp. infection consider raising their water heater temperature to 130 °F (55 °C).

A trial to test whether raising water heater temperature was conducted in homes of older, taller, slender women with *M. avium* lung disease in Wynnewood, Montgomery County, Pennsylvania. Within 8–12 weeks of raising the water heater temperature, *M. avium* in samples of their premise plumbing disappeared (Lande et al., 2019).

Those patient homes have been continually monitored as it would be possible that the elevated water temperature could select for spontaneous temperature-resistant mutants. Rather than discover temperature-resistant mutants, it was discovered that growth of *M. avium* at its growth temperature maximum of 42 °C led to a significant increase in survival after exposure to very high temperatures (Guenette et al., 2020). Rather than being completely killed (three orders of magnitude) upon exposure to 65 °C for 1 h, *M. avium* cells grown at 42 °C exhibited between 30% and 40% survival (Guenette et al., 2020). That increase in thermal tolerance was not due to mutation, but represented a temporary adaptation as cells grown at 25 °C after being grown at 42 °C, regained their susceptibility to exposure to 65 °C (Guenette et al., 2020). Concomitantly, the temperature-tolerant cells of *M. avium* had significantly higher levels of trehalose (Guenette et al., 2020) and trehalose has been implicated in thermal tolerance in other microorganisms.

6.2.6 Desiccation Tolerance

Upon learning of cases of *M. chimaera* infection associated with heater-coolers, one question emerged. How did *M. chimaera* survive in heater-coolers that were assembled in Munich, Germany, then filled with water for testing, emptied, and then shipped throughout the world? One possible explanation was that the cells survived in biofilms on the surface of the pipes, pumps, and reservoirs. We have grown biofilms of *M. chimaera* on stainless steel coupons and measured the survival of colony-forming units over time while under desiccation. Survival was quite high, with 40–60% of the initial biofilm count surviving for as long as 42 days (7 weeks) (Guenette et al., 2020). Thus, a substantial proportion of *M. chimaera* cells would survive to inoculate the water that was added to the heater-cooler before use. Although I have used the example of *M. avium*, desiccation tolerance is characteristic of other OPPPs, in particular *Methylobacterium* spp. (Hugenholtz et al., 1995; Kelley et al., 2004; La Duc et al., 2007; Simmon et al., 2009; Mora et al., 2016).

6.2.7 OPPP Survival and Growth in Amoebae

OPPPs are generally shown to be amoeba-resisting microorganisms (ARM). Rather than being killed following ingestion by amoebae, the OPPPs survive the exposure to amoeba-killing activities and grow. Long-term survival of *L. pneumophila* was observed in *Acanthamoeba castellanii* (Bouyer et al., 2002). In some instances, OPPPs kill the host amoebae, releasing more OPPPs. The number of different free-living amoebae species and ARMs illustrates that the interaction between waterborne bacterial pathogens and amoeba is not unusual (Thomas et al., 2006; Thomas et al., 2008; Lau and Ashbolt, 2009). Resistance to phagocytic killing by free-living amoebae is also a shared characteristic of OPPPs. *L. pneumophila* and *M. avium* are not killed, but survive and grow in phagocytic, free-living amoebae [e.g., *Acanthamoeba, Vermamoeba* (nee, *Hartmanella*)]; they are ARMs (Thomas et al., 2008).

The growth of OPPPs in amoeba can also endow the phagocytized cells with new or modified characteristics. Amoebae-grown *L. pneumophila* (Brieland et al., 1997) and *M. avium* (Cirillo et al., 1997) are more virulent than cells grown in a laboratory medium. In contrast, phagocytosis of *P. aeruginosa* results in the rapid killing of *Acanthamoeba castellanii* (Matz et al., 2008).

One question remains concerning the interaction of OPPPs and amoebae. That is, whether the residence of OPPPs in amoebae is required for virulence? I have hoped to devise a method to identify a sign, a "fingerprint" so-to-speak, of intra-amoebal growth of OPPPs to distinguish those that have grown in amoebae from those that have not to determine whether intra-amoebal growth is required for virulence.

6.2.8 Adaptation to Novel Environmental Stress

Soil-borne and waterborne microorganisms like the OPPPs would be expected to encounter environmental stresses such as low-oxygen concentrations, elevated temperatures, and increased salt (NaCl) concentrations. Rather than rely on low-frequency events such as mutations to ensure survival, the ability to adapt would be a strong selective advantage. Although the data is limited, I suggest that the OPPPs share the characteristic of robust, adaptive stress responses. As is the case throughout this compilation, my model is *M. avium*. The slow growth, coupled with a rapid metabolism suggests that upon encountering stressful conditions (e.g., increasing NaCl concentrations in an estuary during daily high tide events), *M. avium* has time to induce protective proteins and enzymes whose appearance protects against the stress. Above, I related the apparent adaptation of *M. avium* to survival at high temperatures, specifically 60 °C and 65 °C. Although yet unproven, the increased concentrations of membrane-associated trehalose suggests that *M. avium* is able to upregulate trehalose synthesis in the face of changes in growth temperature (Falkinham, unpublished). Such adaptation is a likely survival scenario for the relatively slower-growing OPPPs, as the cells can induce the production of protective proteins (e.g., chaperonins) or compounds (e.g., trehalose) before the cells are killed due to imbalances in growth. I suggest that slow rates of replication (e.g., *M. avium*) may have unexpected benefits to microbial survival.

6.3 Selection for OPPPs

It is important to consider drinking water treatment plants, distribution systems, and premise plumbing as selective, reducing population diversity and allowing the amplification of numbers of a smaller group of microorganisms; here the focus is on opportunistic pathogens. This brief review will document that water systems have been selected for the survival, persistence, and growth of OPPPs.

6.3.1 Selection for Disinfectant Resistance

First, all OPPPs are relatively resistant to chlorine and other disinfectants used in water treatment. That has been shown for *L. pneumophila* (Kuchta et al., 1985), *M. avium* (Taylor et al., 2000), and *P. aeruginosa* (Grobe et al., 2001). The consequence is that OPPPs survive exposure to residual disinfectant levels while competitors for nutrients are killed. *M. avium* is clearly the most resistant to chlorine and only *P. aeruginosa* can grow under anaerobic conditions. However, the levels for all are high enough for them to survive conditions in drinking water distribution systems and premise plumbing.

6.3.2 Selection in Estuaries

Estuaries yield very high numbers of members of the *M. avium* complex (Falkinham et al, 1980), and *P. aeruginosa* has been recovered from estuaries (Diaz et al., 2018). *M. avium* complex cells grow in fresh and brackish (1–2% NaCl), but not in ocean water (George et al., 1980). That provides evidence of another adaptation of *Mycobacterium* spp. to variable environmental conditions, namely low tides with freshwater inputs and high tides with high salt inputs.

6.3.3 Selection Due to Slow Growth

Compared to *Escherichia coli* and other fecal coliforms, the OPPPs grow relatively slowly. However, slow growth is not a detriment to survival, persistence, and growth in premise plumbing. Slow growth means that these opportunistic pathogens are often missed; for example, it takes up to 7 days at 30 °C for the appearance of *Methylobacterium* spp. colonies and up to 14 days for the first appearance of *M. avium* colonies incubated at 37 °C. Slow growth does not represent slow metabolism, just a slow rate of production of daughter cells. Slow growth is actually an advantage because it is associated with slow death. For example, chlorine resistance of *L. pneumophila* and *M. avium* is increased if cells are adapted to drinking water (Taylor et al., 2000; Kuchta et al., 1985).

6.3.4 Surface Adherence and Biofilm Formation

Surface adherence and biofilm formation are of critical selective advantage in a flowing system such as premise plumbing. Adherent microorganisms will not be washed out because of water flow. Further, microorganisms in biofilms are more resistant to antimicrobial agents compared to those in suspension, specifically shown for *M. avium* (Steed and Falkinham, 2006), *L. pneumophila* (Schofield and Locci, 1985), and *P. aeruginosa* (de Beer et al., 1994). The layers of extracellular matrix materials, such as polysaccharide, lipid, DNA, and protein, serve as a barrier to the entry of agents, many of which are hydrophilic (e.g., chlorine).

6.3.5 OPPPs are ARMs

As free-living, phagocytic amoeba and protozoa are in natural and human-engineered water systems, they share nutrients, oxygen, and space with OPPPs. Amoeba and protozoa graze on microorganisms in drinking water systems and premise plumbing, so it follows that there is a survival benefit to the ability of an OPPP to survive and even grow within amoeba and protozoa. As noted in the discussion of OPPPs as ARMs (Chapter 5), the intracellular OPPPs might provide a benefit to the amoebae or protozoa. *T. pyriformis* with intracellular *M. avium* can grow from low-inoculum density, unlike cells lacking intracellular *M. avium* (Falkinham unpublished). Thus, the benefits of phagocytosis of OPPPs may be beneficial to both partners.

6.3.6 Thermal Tolerance

The adaptive thermal tolerance of *M. avium* may not be due to the selection first in premise plumbing, but could be a consequence of the growth of *M. avium* in estuaries. *M. avium* complex numbers in estuaries, such as the Chesapeake and Delaware Bays on the east coast of the United States, are quite high (Falkinham et al., 1980). During the summer months, water temperatures in the shallow, humic, and fulvic acid-rich estuaries can be as high as 45 °C (Kirschner et al., 1992). The humic and fulvic acids, breakdown products of dark, non-cellulosic structural components of woody plants, impart a dark color to estuarine water. Along with this, long summer days lead to elevated water temperatures.

6.3.7 Growth and Survival under Low Oxygen and Anaerobic Conditions

Although I emphasized the growth of OPPPs in stagnant premise plumbing water, low oxygen and anaerobic conditions are met in the natural environment. Habitats that are rich in *Mycobacterium* spp. include peat bogs (Kazda, 2000) and eastern coastal black water acidic swamps of the United States (Kirschner et al., 1992). A major component of a peat bog is incompletely degraded peat that is rich in humic and fulvic acids. Commercial peat-rich potting soil has approximately 1 million *Mycobacterium* spp. cells/gm (de Groot et al., 1999). The partial degradation is due, in part, to the absence of oxygen. There is an insufficient movement of water in peat bogs to oxygenate the water and solid materials. Thus, *Mycobacterium* spp. whose growth is stimulated by humic and fulvic acids (Kirschner et al., 1999), must metabolize and grow at low-oxygen levels (i.e., stagnation).

6.4 Sources of OPPPs

6.4.1 Common OPPP Sources

There is a wide variety of OPPP sources in office buildings, healthcare facilities, homes, apartments, and condominiums; in fact, so many that it is the best to consider that humans are surrounded by OPPPs (Table 6.2). The sources include drinking water, showerheads (Falkinham et al., 2008; Feazel et al., 2009), water taps and water tap aerators (Falkinham, 2011), hot tubs (Katsuda et al., 2018), ice and water from refrigerators, water heaters (Lande et al., 2019), filters, and humidifiers (Hamilton and Falkinham, 2018). Recirculating hot water systems in high-rise condominiums and apartment buildings have especially high numbers of *Mycobacterium* spp. (Tichenor et al., 2012). In hospitals, both filters (Falkinham, 2010) and disinfectant solutions (Weber et al., 2007) have been shown to carry OPPPs, which can then be transmitted to patients.

Table 6.2 Sources of OPPPs

Drinking Water
Showerheads
Water Taps
Water Tap Aerators
Sink Drains
Refrigerator Ice and Water
Water Heaters
Humidifiers
Medical Solutions
Medical Equipment

6.4.2 Novel OPPP Sources

6.4.2.1 Granular-activated charcoal (carbon) filters as OPPP sources

Another source of OPPP infection in homes and hospitals is granular-activated charcoal (carbon) filters. Although called filters, they do not prevent the passage of OPPPs, as the size of the pores is large enough to allow bacteria and OPPPs to pass. Further, GAC filters trap organic matter and metals that adherent OPPPs to grow. A study of a GAC filter showed that the numbers of *M. avium* and other *Mycobacterium* spp. increased following their addition (i.e., inoculation) via a water suspension (Rodgers et al., 1999). Growth was made evident by the repeated recovery of water samples passed through the filter (Rodgers et al., 1999). Unfortunately, a manufacturer's direction for changing the GAC filters is based on its capacity for binding chlorine and metals that impart a bad taste to water. That duration (usually 30 days) is much longer than the time that the EPA investigators demonstrated the presence of *M. avium* in the effluent water (Rodgers et al., 1999).

6.4.2.2 Refrigerator water and ice as OPPP sources

The thoughtful son of parents both with *Mycobacterium abscessus* pulmonary disease led my lab to discover that water and ice from refrigerators were a source of infection. A sample of water from the refrigerator tap had almost 100,000 colony-forming units of *M. abscessus*. The son told us that his mother drank six 8-ounce glasses of water from that tap daily. Upon being told of the results of the analysis of the refrigerator tap, the son sent us the part of the refrigerator involved in the water and ice system. We found it was heavily colonized by *M. abscessus*. As the water system gets heated as the refrigerator cools the interior, water for the refrigerator tap is heated, thus stimulating growth.

6.4.2.3 Dental units and water as OPPP sources

Recent reports in the scientific literature have documented the presence of *L. pneumonia* (Ditommaso et al., 2016) or *M. abscessus* (Hatzenbuehlers et al., 2017; Singh et al., 2018) in dental units and dental unit waterlines. There have been two outbreaks of *M. abscessus* infection in children (i.e., cervical lymphadenitis) reported from dental clinics in Atlanta, GA (Hatzenbuehler et al., 2017) and Anaheim, CA (Singh et al., 2018) performing pulpotomies to remove decayed baby teeth without damaging the permanent tooth below. In light of the presence of OPPPs in drinking water and their resistance to disinfection and ability to form biofilms, those reports should not surprise. As the effectiveness of disinfection protocols is based on *Escherichia coli* as an indicator, those may be inadequate to eradicate OPPPs from dental units.

6.4.2.4 Operating room heater-coolers as OPPP sources

In 2015, the medical world learned of cardiac surgery-linked *Mycobacterium chimaera* infections linked to the use of a single manufacturer's heater-cooler (Sax et al., 2015). Heater-coolers are used to cool patients and maintain blood temperature in conjunction with devices that allow oxygen–carbon dioxide exchange while the patient's heart is halted for repair or replacement. Sadly, at the manufacturing facility, tap water was added to the heater-coolers to test and certify their operation. The facility's water had *M. chimaera,* leading to the colonization of heater-coolers shipped throughout the world. During the use of the heater-cooler, *M. chimaera*-containing aerosols were released from the heater-cooler that infected patients. Although the frequency of the infections was not high, the mortality was 50% and the appearance of symptoms of the *M. chimaera* infection could be quite later than surgery (e.g., up to four years). As a result of those outbreaks, heater-coolers require recertification for use by the United States Food and Drug Administration.

6.5 Transmission of OPPPs

Transmission of OPPPs to individuals occurs most commonly through aerosols. OPPP-laden aerosols (i.e., mists) are generated in showers by *L. pneumophila* (Bollin et al., 1985) and *M. avium* (Falkinham et al., 2008), by aerators attached to water taps, and by splashing as water goes down the drain. Transmission may not only be from the environment, but an interesting hypothesis of transmission of *P. aeruginosa* patients to drains has been proposed (Reuter et al., 2002). As water is employed to humidify water in heating ventilation air conditioning (HVAC), waterborne OPPPs can be aerosolized and circulated throughout a residence, building, or hospital. Further, portable humidifiers, especially those described as ultrasonic, can generate aerosols containing high numbers of OPPPs (Hamilton and Falkinham, 2018).

OPPPs are particularly prone to be aerosolized as their surface is relatively hydrophobic. Hydrophobicity is a driving source for aerosolization, the process of transferring bacteria from water to air by ejection of small droplets (Parker et al., 1983). Following our report of *M. avium* pulmonary disease due to infection by *M. avium* in a patient's showerhead (Falkinham et al., 2008), Dr. Norman Pace of the University of Colorado sent his students across the United States to collect showerhead samples. By isolating and amplifying DNA, they were able to measure the number of gene copies in each showerhead library. Approximately 70% of the showerhead samples had a predominant proportion of *Mycobacterium* spp. sequences and, of those, 30% were *M. avium* (Feazel et al., 2009).

Exposure to water containing OPPPs can also lead to infection. It has been proposed that gastric reflux disease (GERD) could be a risk factor for *M. avium* complex pulmonary disease by aspiration of water or food containing OPPPs (Thomson et al., 2007). For example, fishermen and individuals looking after aquaria for pleasure or business can have cuts infected with *M. avium*. The fact that skin-disinfecting solutions used in hospitals and healthcare facilities can be colonized by OPPPs, leads to infections (Weber et al., 2007).

References

Alary M, Joly JR. 1991. Risk factors for contamination of domestic hot water systems by *Legionellae*. *Appl Environ Microbiol.* 57: 2360–2367.

Bollin GE, Plouffe JF, Para MF, Hackman B. 1985. Aerosols containing *Legionella pneumophila* generated by shower heads and hot-water faucets. *Appl Environ Microbiol.* 50: 1128–1131.

Bouyer S, Imbert C, Rodier M-H, Yann H. 2007. Long-term survival of *Legionella pneumophila* associated with *Acanthamoeba castellanii* vesicles. *Environ. Microbiol.* 9: 1341–1344.

Brennan PJ, Nikaido H. (1995). The envelope of mycobacteria. *Annu Rev Biochem.* 64: 29–63.

Brieland JK, Fantone JC, Remick DG, LeGendre M, McClain M, Engleberg NC. 1997. The role of *Legionella pneumophila*-infected *Hartmanella vermiformis* as an infectious particle in a murine model of Legionnaires' disease. *Infect Immun.* 65: 5330–5333.

Brooks RW, Parker BC, Gruft H, Falkinham JO III. 1984. Epidemiology of infection by nontuberculous mycobacteria. V. Numbers in eastern United States soils and correlation with soil characteristics. *Am. Rev. Respir. Dis.* 130: 630–633.

Ciesielski CA, Blaser MJ, Wang W-LL. 1984. Role of stagnation and obstruction of water flow in isolation of *Legionella pneumophila* from hospital plumbing. *Appl Environ Microbiol.* 48: 984–987.

Cirillo JD, Falkow S, Tompkins LS, Bermudez LE. 1997. Interaction of *Mycobacterium avium* with environmental amoebae enhances virulence. *Infect Immun.* 65: 3759–3767.

Cooper IR, Hanlon GW. 2010. Resistance of *Legionella pneumophila* serotype 1 biofilms to chlorine-based disinfection. *J Hosp Infect.* 74: 152–159.

De Beer D, Srinivasan R, Stewart PS. 1994. Direct measurement of chlorine penetration into biofilms during disinfection. *Appl Environ Microbiol.* 60: 4339–4344.

Diaz KE, Remold SK, Onyiri O, Bozeman M, Raymond PA, Turner PE. 2018. Generalized growth of estuarine, household and clinical isolates of *Pseudomonas aeruginosa*. *Front Microbiol.* 9: 305.

Ditommaso S, Giacomuzzi M, Ricciadi E, Zotti CM. 2016. Cultural and molecular evidence of *Legionella* spp. colonization in dental unit waterlines: Which is the best method for risk assessment? *Int J Environ Res Publ Hlth.* 13: 211.

Falkinham JO III, Iseman MD, de Haas P, van Soolingen D. 2008. *Mycobacterium avium* in a shower linked to pulmonary disease. *J Water Health.* 6: 209–213.

Falkinham JO III, Parker BC, Gruft H. 1980. Epidemiology of infection by nontuberculous mycobacteria. I. Geographic distribution in the eastern United States. *Am Rev Respir Dis.* 121: 931–937.

Falkinham JO III. 2010. Hospital water filters as a source of *Mycobacterium avium* complex. *J Med Microbiol.* 59: 1198–1202.

Falkinham JO III. 2011. Nontuberculous mycobacteria from household plumbing of patients with nontuberculous mycobacterial disease. *Emerg Infect Dis.* 17: 419–424.

Falkinham JO III. 2020. Disinfection and cleaning of heater-cooler units: Suspension- and biofilm-killing. *J. Hosp. Infection*. 105: 552–557.

Favero MS, Carson LA, Bond WW, Petersen NJ. 1971. *Pseudomonas aeruginosa*: Growth in distilled water from hospitals. *Science*. 175: 836–838.

Feazel LM, Baumgartner LK, Peterson KL, Frank DN, Harris JK, Pace NR. 2009. Opportunistic pathogens enriched in showerhead biofilms. *Proc Natl Acad Sci USA*. 106: 16393–16399.

George KL, Parker BC, Gruft H, Falkinham JO III. 1980. Epidemiology of infection by nontuberculous mycobacteria. II. Growth and survival in natural waters. *Am Rev Respir Dis*. 122: 89–94.

Grobe S, Wingender JW, Flemming H-C. 2001. Capability of mucoid *Pseudomonas aeruginosa* to survive in chlorinated water. *Int J Hyg Environ Hlth*. 204: 139–142.

Guenette S, Williams MD, Falkinham JO III. 2020. Growth temperature, trehalose, and susceptibility to heat in *Mycobacterium avium*. *Pathogens*. 9: 657.

Hamilton L, Falkinham JO III. 2018. Aerosolization of *Mycobacterium avium* and *Mycobacterium abscessus* from a household ultrasonic humidifier. *J. Med. Microbiol*. 67: 1491–1497.

Hatzenbuehlore LA, Tobin D'Angelo M, Drenzek C, Peralta G, Cranmer LC, Anderson EJ, Milla SS, Abramowica S, Yi J, ilinski J, Rajan R, Whitley MK, Gower V, Berkowitz, Shapiro CA, Williams JK, Harmon P, Shane AL. 2017. Pediatric dental clinic-associated outbreak of Mycobacterium abscessus infection. *J Ped Infect Dis Soc*. 6: e116–e22.

Hugenholtz P, Cunningham MA, Hendrlkz JK, Fuerst JA. 1995. Desiccation resistance of bacteria isolated from an air-handling system biofilm determined using a simple quantitative membrane filter method. *Lttrs Appl Microbiol*. 21: 41–46.

Katsuda R, Yoshida S, Tsuyuguchi K, Kawamura T. 2018. A case report of hot tub lung: Identical strains of *Mycobacterium avium* from the patient and the bathroom air. *Int J Tubercul Lung Dis*. 22: 3520352.0.000822.

Kazda J. 2000. *The ecology of mycobacteria*. Kluwer Academic Publishers, Dordrecht, the Netherlands.

Kelley ST, Theisen U, Angenent LT, St Amand A, Pace NR. 2004. Molecular analysis of shower curtain biofilm microbes. *Appl Environ Microbiol*. 70: 4187–4192.

Kirschner RA Jr, Parker BC, Falkinham JO III. 1992. Epidemiology of infection by nontuberculous mycobacteria. X. *Mycobacterium avium, M. intracellulare*, and *M. scrofulaceum* in acid, brown-water swamps of the southeastern United States and their association with environmental variables. *Am Rev Respir Dis.* 145: 271–275.

Kirschner RA Jr, Parker BC, Falkinham JO III. 1999. Humic and fulvic acids stimulate the growth of *Mycobacterium avium. FEMS Microbiol Ecol.* 30: 327–332.

Kuchta JM, States SJ, McNamara AM, Wadowsky RM, Yee RB. 1983. Susceptibility of *Legionella pneumophila* to chlorine in tap water. *Appl Environ Microbiol.* 46: 1134–1139.

La Duc MT, Dekas A, Osman S, Moissl C, Newcombe D, Venkateswaran K. 2007. Isolation and characterization of bacteria capable of tolerating the extreme conditions of clean room environments. *Appl Environ Microbiol.* 73: 2600–2611.

Lande L, Alexander DC, Wallace RJ Jr, Kwait R, Iakhiaeva E, Williams M, Cameron ADS, Olshefsky S, Devon R, Vasireddy R, Peterson DD, Falkinham JO III. 2019. *Mycobacterium avium* in community and household water, suburban Philadelphia, Pennsylvania, USA, 2010–2012. *Emerg Infect Dis.* 25: 473–481.

Lau HY, Ashbolt NJ. 2009. The role of biofilms and protozoa in *Legionella* pathogenesis: Implications for drinking water. *J Appl Microbiol.* 107: 368–378.

Lewis AH, Falkinham JO III. 2015. Microaerobic growth and anaerobic survival of *Mycobacterium avium, Mycobacterium intracellulare*, and *Mycobacterium scrofulaceum. Int J Mycobacteriol.* 4: 25–30.

Matz C, Moreno AM, Alhede M, Manefield M, Hauser AR, Givskov M, Kjelleberg S. 2008. *Pseudomonas aeruginosa* uses type III secretion system to kill biofilm-associated amoebae. *ISME J.* 2: 843–852.

Mauchline JWS, Araujo R, Wait R, Dowsett AB, Dennis PJ, Keevil CW. 1992. Physiology and morphology of *Legionella pneumophila* in continuous culture at low oxygen concentration. *J gen Microbiol.* 138: 2371–2380.

Mora M, Perras A, Alekhova TA, Wink L, Krause R, Aleksandrova A, Novozhilova, Moissl-Eichinger C. 2016. Resilient microorganisms in dust samples of the International Space Station-survival of the adaptation specialists. *BioMed Central.* 4: 65.

Mullis SN, Falkinham JO III. 2013. Adherence and biofilm formation of *Mycobacterium avium* and *Mycobacterium abscessus* to household plumbing materials. *J Appl Microbiol.* 115: 908–914.

Norton CD, LeChevallier MW, Falkinham JO III. 2004. Survival of *Mycobacterium avium* in a model distribution system. *Water Res.* 38: 1457–1466.

Palmer KL, Brown SA, Whitley M. 2007. Membrane-bound nitrate reductase is required for anaerobic growth in cystic fibrosis sputum. *J Bacteriol.* 189: 4449–4455.

Parker BC, Ford MA, Gruft H, Falkinham JO III. 1983. Epidemiology of infection by nontuberculous mycobacteria. IV. Preferential aerosolization of *Mycobacterium intracellulare* from natural waters. *Am. Rev. Respir. Dis.* 128: 652–656.

Paszko-Kolva C, Shahamat M, Colwell RR. 1992. Long-term survival of *Legionella pneumophila* serogroup 1 under low-nutrient conditions and associated morphological changes. *FEMS Microbiol. Ecol.* 11: 45–55.

Reuter S, Sigge A, Wiedeck H, Trautmann M. 2002. Analysis of transmission pathways of *Pseudomonas aeruginosa* between patients and tap water outlets. *Crit Care Med.* 30: 2222–2228.

Rodgers MR, Blackstone BJ, Reyes AL, Covert TC. 1999. Colonization of point of use water filters by silver resistant non-tuberculosis mycobacteria. *J Clin Pathol.* 52: 629.

Sax H, Bloemberg G, Hasse B, Sommerstein R, Kohler P, Achermann Y, Rössle M, Falk V, Kuster SP, Böttger EC, Weber R. 2015. Prolonged outbreak of *Mycobacterium chimaera* infection after open-chest heart surgery. *Clin Infect Dis.* 61: 67–75.

Schofield GM, Locci R. 1985. Colonization of components of a model hot water system by *Legionella pneumophila. J appl Bacteriol.* 58: 151–162.

Simmon RB, Rose LJ, Crow SA, Ahearn DG. 1999. The occurrence and persistence of mixed biofilms in automobile air conditioning systems. *Current Microbiol.* 39: 141–145.

Singh J, O'Donnell K, Ashouri N, Adler-Shohet FC, Nieves D, Tran MT, Arrieta A. 2018. Outbreak of invasive nontuberculous Mycobacterium (NTM) infections associated with a pediatric dental practice, *Open Forum Infectious Diseases.* 5: S29; https://doi.org/10.1093/ofid/ofy209.067.

Steed KA, Falkinham JO III. 2006. Effect of growth in biofilms on chlorine susceptibility of *Mycobacterium avium* and *Mycobacterium intracellulare. Appl Environ Microbiol.* 72: 4007–4010.

Straus WL, Plouffe JF, File TM Jr, Lipman HB, Hackman BH, Salstrom S-J, Benson RF, Breiman RF. 1996. Risk factors for domestic acquisition of Legionnaires Disease. *Arch Intern Med.* 156: 1685–1692.

Taylor RM, Norton CD, LeChevallier MW, Falkinham JO III. 2000. Susceptibility of *Mycobacterium avium, Mycobacterium intracellulare,* and *Mycobacterium scrofulaceum* to chlorine, chloramine, chlorine dioxide, and ozone. *Appl Environ Microbiol.* 66: 1702–1705.

Thomas V, Loret J-F, Jousset M, Greub G. 2008. Biodiversity of amoebae and amoebae-resisting bacteria in a drinking water treatment plant. *Environ Microbiol.* 10: 2728–2745.

Thomas V, Herrera-Rimann K, Blanc DS, Greub G. 2006. Biodiversity of amoebae and amoebae-resisting bacteria in a hospital water network. *Appl. Environ. Microbiol.* 72: 2428–2438.

Thomson RM, Armstrong JG, Looke DF. 2007. Gastroesophageal reflux disease, acid suppression, and *Mycobacterium avium* complex pulmonary disease. *Chest.* 131: 1166–1172.

Tichenor WS, Thurlow J, McNulty S, Brown-Elliott BA, Wallace RJ Jr, Falkinham JO III. 2012. Non-tuberculosis mycobacteria in household plumbing as possible cause of chronic rhinosinusitis. *Emerg Inf Dis.* 18: 1612–1617.

Van Loosdrecht MCM, Lyklema J, Norde W, Zehnder AJB. 1989. Bacterial adhesion: A physicochemical approach. *Microb Ecol.* 17: 1–15.

Weber DJ, Rutala WA, Sickbert-Bennett EE. 2007. Outbreaks associated with contaminated antiseptics and disinfectants. *Antimicrob. Agents Chemother.* 51: 4217–4224.

Chapter 7

Introduction to Emerging Opportunistic Premise Plumbing Pathogens

In addition to listing and describing the salient characteristics of accepted opportunistic premise plumbing pathogens(OPPPs), the following sections introduce and discuss a group of emerging OPPPs. As examples of emerging OPPPs, I include *Stenotrophomonas maltophilia, Acinetobacter baumannii, Sphingomonas paucimobilis, Methylobacterium* spp., *Segniliparus* spp., and *Cupriavidus* spp. As noted for the chapter introducing the established OPPPs (Chapter 4), this group of bacteria is included to serve as models for OPPPs and, further, to identify unique characteristics of these emerging OPPPs to expand understanding of the range of microorganisms that are OPPPs. An example of the novel features is the desiccation resistance of *Methylobacterium* spp. that is a likely selective factor in premise plumbing that suffers from intermittent water flow.

The following sections (i.e., Sections 7.1 through 7.6) provide brief reviews of the infections, physiology, and sources of emerging OPPPs, as known at present. Thus, the information is limited and the sections are relatively short. I hope that the inclusion of these emerging OPPPs will stimulate further research and study of these bacteria and enrich the knowledge of the entire group of waterborne opportunistic pathogens.

Opportunistic Premise Plumbing Pathogens
Joseph O. Falkinham, III

ISBN 978-981-4968-40-9 (Hardcover), 978-1-003-32100-2 (eBook)
www.jennystanford.com

7.1 *Stenotrophomonas maltophilia*

7.1.1 Diseases and Risk Factors of *S. maltophilia*

Stenotrophomonas maltophilia, first included as a member of the genus *Pseudomonas*, then transferred to *Xanthomonas*, is a Gram-negative, waterborne, emerging opportunistic pathogen (Looney et al., 2009). It is primarily associated with hospital-acquired infections, primarily bacteremia associated with exposure of individuals of increased susceptibility to infection including neonates (Verweij et al., 1998) and patients with cancer (Garcia Paez and Costa, 2008), kidney disease requiring hemodialysis (Arvanitidou et al., 2003), cystic fibrosis (Stanojevic et al., 2013), and as a consequence of surgery, including cataract surgery (Chang et al., 2013). Risk factors for *S. maltophilia* infection include residence in an intensive care unit (ICU), presence of a central venous catheter (CVC), hematologic malignancy, shock, and thrombocytopenia (Garcia Paez and Costa, 2008). In another study, chronic obstructive pulmonary disease (COPD) and long duration of antibiotic therapy were independent risks for *S. maltophilia* infections acquired in an ICU (Nseir et al., 2006). *S. maltophilia* causes a variety of ocular infections (Watanabe et al., 2014). In particular, a proportion of contact lens wearers are at risk for keratitis with an incidence of 2.7–4.2 per 100,000 per year (Watanabe et al., 2014).

Infections caused by *S. maltophilia* are relatively rare, and although there have been a few studies documenting their prevalence or incidence, infections are increasing. Over the period 1985–2008, 38 cases of *S. maltophilia* infection with risk factors associated with mortality were identified by searching PubMed and OVID databases (Garcia Paez and Costa, 2008). The prevalence of *S. maltophilia* pulmonary infection in cystic fibrosis patients is evidently increasing. Over the period, August 2006 through June 2008, 267 cystic fibrosis patients were studied and the prevalence of *S. maltophilia* in sputum cultures increased from 6.7% to 12.0% of patients. Finally, as strains of *S. maltophilia* are used for biological control of plant pathogens and remediation (Berg et al., 1999), their presence in the environment may increase infection incidence and prevalence.

7.1.2 Ecology and Habitats of *S. maltophilia*

S. maltophilia has been isolated from a variety of environmental sites, including salt-rich soil (Alexander et al., 2020), deionized water (Wishart and Riley, 1976), tap water (Verweij et al., 1998; Brooke, 2012), water from oxygen humidifier reservoirs (Cameron et al., 1986), water collected from hemodialysis centers (Arvanitidou et al., 2003), faucet aerators (Weber et al., 1999), dental unit waterlines (Szymańska and Sitkowska, 2013), and heater-coolers (Chand et al., 2017). Those publications primarily report isolation from hospitals and it would be understood that *S. maltophilia* is also present in plumbing in households, apartments, and condominiums. A study of *S. maltophilia* presence in clinics, hospitals, and homes of cystic fibrosis patients revealed that 67 of 194 (28%) samples from water, faucets, sink drains, and other samples yielded *S. maltophilia* (Denton et al., 1998). Interestingly, the percentage of *S. maltophilia* recovery from hospital and clinic samples (28%) was lower than that from homes (39%) (Denton et al., 1998), suggesting that premise plumbing is an ideal habitat for *S. maltophilia*.

7.1.3 Source Tracking and Fingerprinting of *S. maltophilia*

Water has been the source of infection that cannot simply rely upon isolation of *S. maltophilia* and water to which a patient was exposed, a DNA-based typing method must be used. A variety of typing methods have been employed in the search for environmental sources of *S. maltophilia* or for identifying types of higher or lower "virulence" (Gherardi et al., 2015). In that work, "virulence" was defined as an *S. maltophilia* type that is more common in patients compared to environmental isolates. Methods for typing *S. maltophilia* include pulsed-field gel electrophoresis (PFGE), randomly amplified polymorphic DNA (RAPD), restriction fragment length polymorphism (RFLP) analysis of specific genes (e.g., *gyrB*), enterobacterial repetitive intergenic sequences (ERIC or *rep*-PCR), multilocus variable-number tandem repeat analysis (MLVA), and multilocus sequence typing (MLST) (Gherardi et al., 2015). Some of the methods have drawbacks, e.g., PFGE is time-consuming and RAPD typing is difficult to replicate. Further, care must be taken in choosing a method that provides a sufficient level of discrimination; too much and all strains are unique and too little discrimination and all strains are part of the same clone. Two studies have shown that PFGE was more discriminatory than *rep*-PCR, as isolates that shared the same *rep*-PCR type might have different PFGE patterns (Denton et al., 1998; Berg et al., 1999). A comparison of *rep*-PCR and partial *gyrB*-gene sequencing displayed that *rep*-PCR identified 12 subgroups and *gyrB*-sequencing 10, showing the two methods provided approximately the same level of discrimination with *rep*-PCR slightly better (Adamek et al., 2011). MLST may lack discrimination although it is valuable for the identification of regional geographic clones or virulence types (Gherardi et al., 2015). Finally, another consideration for the choice of typing method is the ability to compare data from different studies in different laboratories throughout the world. Here, MLVA typing is superior as it provides a numerical type, for each locus the variable number of repeats are reported and the number of repeats at each locus analyzed can be compared between strains.

Proof of the environmental origin of *S. maltophilia* was presented in a study of 139 *S. maltophilia* isolates recovered from a single hospital over a four-year period (Valdezate et al., 2004). An isolate obtained from a bronchoscope had the same PFGE pattern as did patient isolates exposed to that bronchoscope (Valdezate et al., 2004). However, it was possible that the bronchoscope was contaminated by a patient sample and thereby transmitted to the remaining patients. In another study of 41 cystic fibrosis patients, isolates from patients and their environment (both home and hospital) were typed using the *rep*-PCR and PFGE. Thirty-three of the 41 patients (80%) were colonized with unique strains of *S. maltophilia* (Denton et al., 1998). Although there was no evidence of patient-to-patient transmission, evidence showed that two patients were infected with a strain that shared the same fingerprint as isolates recovered from the clinic or hospital environment. None of the patients were infected with a strain that was isolated from their homes (Denton et al., 1998).

The search for virulence-specific genetic signatures has been unsuccessful to date. An amoebae model of virulence assessment has been employed. Different numbers of *Dictyostelium discoideum* or *Acanthamoeba castellanii* were spotted on lawns of *S. maltophilia.* The higher the number of protists required for the formation of a plaque (hole) in the *S. maltophilia* lawn, the higher the virulence of the *S. maltophilia* strain (Adamek et al., 2011). Although a virulence island and putative virulence genes were identified by genome sequences of *S. maltophilia* strains, their presence or absence did not correlate with virulence in the amoebae model (Adamek et al., 2011; Adamek et al., 2014).

7.1.4 Genetic Variability of *S. maltophilia*

S. maltophilia isolates from patients in a single hospital demonstrate a high genetic diversity (Valdezate et al., 2004; Turrientes et al., 2010; Vidigal et al., 2014). That diversity has led to the hypothesis that there is a strong selection for mutator strains (i.e., strains with higher mutation rates) in patients due to the stressful conditions in the infected human lung (Turrientes et al., 2010; Vidigal et al., 2014). Such adaptability is consistent with the presence of potentially mobile regions of DNA in the *S. maltophilia* genome (Crossman et al., 2008) and its versatility as a plant growth stimulator, bioremediation microorganism, as well as a human pathogen (Ryan et al., 2009). Not only *S. maltophilia,* but other OPPPs, such as *P. aeruginosa* and *A. baumannii,* might behave similarly. Consistent with that hypothesis, low-mutation frequencies were found amongst environmental isolates of *S. maltophilia,* whereas hypermutator isolates were only found amongst patient isolates (Turrientes et al., 2010). In a subsequent study of *S. maltophilia* isolates from cystic fibrosis patients, the hypermutator isolates were present immediately after infection but disappeared over time, and were replaced with cells that were less mutable (Vidigal et al., 2014). Perhaps, high-mutation frequencies were adaptive in allowing an establishment of an infecting population of *S. maltophilia* during the initial, stressful adaptation from environmental to patient habitats. Following that period of adaptation, a lower frequency of mutation allows for continued residence in the infected lung.

7.1.5 Physiological Ecology of *S. maltophilia*

Although there are ample publications describing the sources of *S. maltophilia* and infections it causes, there is little on its ecology and those physiological characteristics that influence its presence in drinking water and premise plumbing. Fortunately, a selective isolation and enumeration method has been described (Bollet et al., 1995) that should make surveys of premise plumbing and hospital water systems possible. It is well established that hospital tap water is a source of *S. maltophilia* infection (Cervia et al., 2008). However, its long-term persistence in drinking water and premise plumbing might require the presence of amoebae, as it was observed that 28-day survival in laboratory suspensions required the presence of *Vermamoeba vermiformis* (Cateau et al., 2014). *S. maltophilia* is an epiphyte of many plants (Ryan et al., 2009) and thereby can be transmitted from plant leaf to water. Further, *S. maltophilia* is capable of metabolizing a wide variety of organic compounds, such as phenols, in common with pseudomonads (Ryan et al., 2009).

Although there is just a single published study (Yorioka et al., 2010), it appears that *S. maltophilia* is quite resistant to chlorine. In that experiment, cells of *S. maltophilia* in tube biofilms were exposed to 1,000 ppm sodium hypochlorite and survivors enumerated (Yorioka et al., 2010). Even at that high concentration of hypochlorite, *S. maltophilia* survival was high, and the calculated $CT_{99.9\%}$ value was greater than 1,000. *S. maltophilia* is also quite resistant to disinfectants used in hospitals for skin disinfection, namely chlorhexidine and cetrimide (Wishart and Riley, 1976). The efficacy of copper-silver (Cu/Ag) ionization systems in reducing numbers of *S. maltophilia* has been measured in a model plumbing system (Shih and Lin, 2010). Following inoculation of bacteria, 14 days of incubation occurred to allow for the establishment of the cells in the system and formation of biofilms. Cu/Ag concentrations from 0.2/0.02 to 0.8/0.08 for 72 h were able to reduce colony counts of biofilm *S. maltophilia* by three logs, and planktonic counts by greater than six logs (Shih and Lin, 2010). Following the release from Cu/Ag production, colony counts rebounded, showing that cells were not killed, just inhibited from growing (Shih and Lin, 2010).

S. maltophilia is also resistant to triclosan (2,4,4'-trichloro-2'-hydroxydiphenylether) and detergent sodium dodecyl sulfate (Brooke, 2102). It is also relatively resistant to contact lens disinfecting solutions, such as polyquaternium, alexidine, polyaminopropyl biguanide, and myristamidopropyl dimethylamine (Watanabe et al., 2014), and to peracetic acid and hydrogen peroxide (Sacchetti et al., 2009).

S. maltophilia attaches rapidly to polystyrene (Di Bonaventura et al., 2004) and is recovered from biofilms in premise plumbing (Denton et al., 1998; Weber et al., 1999). Further, *S. maltophilia* forms biofilms over a wide range of temperatures (i.e., 18 °C–37 °C) and pH (i.e., 5.5–8.5) (Di Bonventura et al., 2007). In addition, *S. maltophilia* also forms biofilms under anaerobic conditions (Di Bonaventura et al., 2004). The latter suggests that *S. maltophilia* can survive in biofilms during the stagnation that is a characteristic of premise plumbing. As other OPPPs, *S. maltophilia* grows in *Vermamoeba vermiformis* (nee *Hartmanella vermiformis*) without reducing the viability of the amoebal host and survives for long periods of time (28 days) under conditions where the bacteria did not survive without amoebae (Cateau et al., 2014).

One interesting feature of *S. maltophilia* is its ability to form, under starvation or stress, ultra-microbial cells that are capable of passing through microbial filters (i.e., 0.2 μm pore size) (Brooke, 2012). Such cells are viable, can grow in biofilms, and thus serve as a source to colonize other habitats (Brooke, 2012). In addition, ultra-microbial cells have been reported for *Pseudomonas* spp., *Sphingomonas* spp., and *Aeromonas* spp. (Brooke, 2012).

References

Adamek M, Linke B, Schwartz T. 2014. Virulence genes in clinical and environmental *Stenotrophomonas maltophilia* isolates: A genome sequencing and gene expression approach. *Microb Pathog*. 67–68: 30–30.

Adamek M, Overhage J, Bathe S, Winter J, Fischer R, Schwartz T. 2011. Genotyping of environmental and clinical *Stenotrophomonas maltophilia* isolates and their pathogenic potential. *PLoS ONE.* 6: e27615.

Alexander A, Singh VK, Misra A. 2020. Halotolerant PGPR *Stenotrophomonas maltophilia* BJ01 induces salt tolerance by modulating physiology and biochemical activities of Arachis hypogaea. *Front Microbiol.* 11: 568289.

Arvanitidou M, Vayona A, Spanakis N, Tsakris A. 2003. Occurrence and antimicrobial resistance of Gram-negative bacteria isolated in haemodialysis water and dialysate of renal units: Results of a Greek multicenter study. *J Appl Microbiol.* 95: 180–185.

Berg F, Roskot N, Smalla K. 1999. Genotypic and phenotypic relationships between clinical and environmental isolates of *Stenotrophomonas maltophilia. J Clin Microb.* 37: 3594–3600.

Bollet C, Davin-Regli A, De Micco P. 1995. A simple method for selective isolation of *Stenotrophomonas maltophilia* from environmental samples. *Appl Environ Microbiol.* 61: 1653–1654.

Brooke JS. 2012. *Stenotrophomonas maltophilia*: An emerging global opportunistic pathogen. *Clin Microbiol Revs.* 25: 2–41.

Cameron JL, Reese WA, Tayal VS, Clark RF, Kelso D, Gonzalez ER, Garnett AR, Ornato JP. 1986. Bacterial contamination of ambulance oxygen humidifier water reservoirs: A potential source of pulmonary infection. *Ann Emerg Med.* 15: 13001302.

Cateau E, Maisonneuve E, Peguilhan S, Quellard N, Hechard Y, Rodier M-H. 2014. *Stenotrophomonas maltophilia* and *Vermamoeba vermiformis* relationships: Bacterial multiplication and protection in amoebal-derived structures. *Res Microbiol.* 165: 847–851.

Cervia JS, Ortolano GA, Canonica FP. 2008. Hospital tap water as a source of *Stenotrophomonas maltophilia* infection. *Clin Infect Dis.* 46: 1485–1487.

Chand M, Lamagni T, Kranzer K, Hedge J, Moore G, Parks S, Collins S, del Ojo Elias C, et al. 2017. Insidious risk of severe *Mycobacterium chimaera* infection in cardiac surgery patients. *Clin Infect Dis.* 64: 335–342.

Chang JS, Flynn HW Jr, Miller D, Smiddy WE. 2013. *Stenotrophomonas maltophilia* endophthalmitis following cataract surgery: Clinical and microbiological results. *Clin Opthal.* 7: 771–777.

Denton M, Todd NJ, Kerr KG, Hawkey PM, Littlewood JM. 1998. Molecular epidemiology of *Stenotrophomonas maltophilia* isolated from clinical specimens from patients with cystic fibrosis and associated environmental samples. *J Clin Microbiol.* 36: 1953–1958.

Di Bonaventura G, Stepanovic S, Picciani C, Ponpilio A, Piccolomini R. 2007. Effect of environmental factors on biofilm formation by clinical *Stenotrophomonas maltophilia* isolates. *Folia Microbiol.* 52: 86–90.

Esposito A, Pompilio A, Bettua C, Crocetta V, Giacobazzi E, Fiscarelli E, Jousson O, Di Bonaventura G. 2017. Evolution of *Stenotrophomonas maltophilia* in cystic fibrosis lung over chronic infection: A genomic and phenotypic population study. *Front Microbiol.* 8: 1590.

Fisher MC, Long SS, Roberts EM, Dunn JM, Balsara RK. 1981. *Pseudomonas maltophilia* bacteremia in children undergoing open heart surgery. *J Am Med Assoc.* 246: 1571–1574.

Garcia Paez JI, Costa ST. 2008. Risk factors associated with mortality of infections caused by *Stenotrophomonas maltophilia*: A systematic review. *J Hosp Infect.* 70: 101–108.

Gherardi G, Creti R, Pompilio A, Di Bonaventura G. 2015. An overview of various typing methods for clinical epidemiology of the emerging pathogen *Stenotrophomonas maltophilia. Diag Microb Infect Dis.* 81: 219–226.

Kampmeier S, Pillukat MH, Pettke A, Kossow A, Idelevich EA, Mellman A. 2017. Evaluation of Stenotrophomonas maltophilia bacteremia cluster in hematopoietic stem cell transplantation recipients using whole genome sequencing. *Antimicrob Rest Infect Contrl.* 6: 115.

Looney WJ, Narita M, Mühlemann K. 2009. *Stenotrophomonas maltophilia*: An emerging opportunist human pathogen. *Lancet Infect Dis.* 9: 312–323.

Morrison AJ Jr., Hoffman KK, Wenzel RP. 1986. Associated mortality and clinical characteristics of nosocomial *Pseudomonas maltophilia* in a university hospital. *J Clin Microbiol.* 24: 52–55.

Mukherjee P, Roy P. 2016. Genomic potential of *Stenotrophomonas maltophilia* in bioremediation with an assessment of its multifaceted role in our environment. *Front Microbiol.* 7: 967.

Nicoletti M, Iacobino A, Prosseda G, Fiscarelli E, Zarrilli R, De Carolis E, Petrucca A, Nencioni L, Colonna B, Casalino M. 2011. *Stenotrophomonas maltophilia* strains from cystic fibrosis patients:

Genomic variability and molecular characterization of some virulence determinants. *Int J Med Microbiol.* 301: 34–43.

Nseir1 S, Di Pompeo C, Brisson H, Dewavrin F, Tissier S, Diarra M, Boulo M, Durocher A. 2006. Intensive care unit-acquired *Stenotrophomonas maltophilia*: Incidence, risk factors, and outcome. *Crit Care.* 10: R143.

Pompilio A, Crocetta V, De Nicola S, Verginelli F, Fiscarelli E, Di Bonaventura G. 2015. Cooperative pathogenicity in cystic fibrosis: *Stenotrophomonas maltophilia* modulates *Pseudomonas aeruginosa* virulence in mixed biofilm. *Front Microbiol.* 6: 951.

Pompilio A, Crocetta V, Ghosh D, Chakrabarti M, Gherardi G, Agostino Vitali L, Fiscarelli E, Di Bonaventura G. 2016. *Stenotrophomonas maltophilia* phenotypic and genotypic diversity during 10-year colonization in the lungs of a cystic fibrosis patient. *Front Microbiol.* 7: 1551.

Pompilio A, Piccolomini R, Picciana C, D'Antonio D, Savini V, Di Bonaventura G. 2008. Factors associated with adherence to and biofilm formation on polystyrene by *Stenotrophomonas maltophilia*: The role of cell surface hydrophobicity and motility. *FEMS Microbiol Lett.* 287: 41–47.

Pompilio A, Pomponio S Crocetta V, Gherardi G, Verginelli F, Fiscarelli, E, Dicuonzo V, Savini V, D'Antonio D, Di Bonaventura G. 2011. Phenotypic and genotypic characterization of *Stenotrophomonas maltophilia* isolates from patients with cystic fibrosis: Genome diversity, biofilm formation, and virulence. *BMC Microbiol.* 11: 159.

Ryan RP, Monchy S, Cardinale M, Taghavi S, Crossman L, Avison MB, Berg G, van der Lelie D, Dow JM. 2009. The versatility and adaptation of bacteria from the genus *Stenotrophomonas. Nature Rev Microbiol.* 7: 514–525.

Sacchett R, De Luca G, Zanetti F. 2009. Control of *Pseudomonas aeruginosa* and *Stenotrophomonas maltophilia* contamination of microfiltered water dispensers with peracetic acid and hydrogen peroxide. *Int J Food Microbiol.* 132: 162–166.

Sakhnini E, Weissmann A, Oren I. 2002. Fulminant *Stenotrophomonas maltophilia* soft tissue infection in immunocompromised patients: An outbreak transmitted via tap water. *Am J Med Sci.* 323: 269–272.

Shih H-Y, Lin YE. 2010. Efficacy of copper-silver ionization in controlling biofilm- and plankton-associated waterborne pathogens. *Appl Environ Microbiol.* 76: 2032–2035.

Stanojevic S, Ratjen F, Stephens D, Lu A, Yau Y, Tullis E, Waters V. 2013. Factors influencing the acquisition of *Stenotrophomonas maltophilia* infection in cystic fibrosis patients. *J Cyst Fibr.* 12: 575–583.

Steinmann J, Mamat U, Abda EM, Kirchhoff L, Streit WR, Schaible UE, Niemann S, Kohl TA. 2018. Analysis of phylogenetic variation of *Stenotrophomonas maltophilia* reveals human-specific branches. *Front Microbiol.* 9: 806.

Szymańska J, Sitkowska J. 2013. Bacterial contamination of dental unit waterlines. *Environ Monit Assess.* 185: 3603–3611.

Turrientes MC, Baquero MR, Sánchez MB, Valdezate S, Escudero E, Berg G, Cantón R, Baquero F, Galaán JC, Martínez JL. 2010. Polymorphic mutation frequencies of clinical and environmental *Stenotrophomonas maltophilia* populations. *Appl Environ Microbiol.* 76: 1746–1758.

Valdezate S, Vindel A, Martín-Dávila P, Sánchez Del Saz B, Baquero F, Cantón R. 2004. High genetic diversity among *Stenotrophomonas maltophilia* strains despite their originating at a single hospital. *J Clin Microbiol.* 42: 693–699.

Verweij PE, Meis JFG, Christmann V, van der Bor M, Melchers WJG, Hilderink BGM, Voss A. 1998. Nosocomial outbreak of colonization and infection with *Stenotrophomonas maltophilia* in preterm infants associated with contaminated tap water. *Epidemiol Infect.* 120: 251–256.

Vidigal PG, Dittmer S, Steinmann E, Buer J, Rath PM, Steinmann J. 2014. Adaptation of *Stenotrophomonas maltophilia* in cystic fibrosis: Molecular diversity, mutation frequency and antibiotic resistance. *Int J Med Microbiol.* 304: 613–619.

Watanabe K, Zhu H, Willcox M. 2014. Susceptibility of *Stenotrophomonas maltophilia* clinical isolates to antibiotics and contact lens multipurpose disinfecting solutions. *Invest Ophthalmol Vis Sci.* 55: 8475–8479.

Weber DJ, Rutala WA, Blanchet CN, Jordan M, Gergen MF. 1999. Faucet aerators: A source of patient colonization with *Stenotrophomonas maltophilia. Am J Infect Control.* 27: 59–63.

Wishart MM, Riley TV. 1976. Infection with *Pseudomonas maltophilia* hospital outbreak due to contaminated disinfectant. *Med J Austral.* 2: 710–712.

Yorioka K, Oie S, Kamiya A. 2010. Microbial contamination of suction tubes attached to suction instruments and preventive methods. *Jpn J Infect Dis.* 63: 124–127.

7.2 *Acinetobacter baumannii*

7.2.1 Introduction

For many water professionals, the announcement of injured U.S. troops in the Middle East with infections caused by *A. baumannii* (Scott et al., 2004) was their first introduction to this emerging opportunistic premise plumbing pathogen. In part, this was due to the long, convoluted history of the taxonomy of the *Acinetobacter* group. *Acinetobacter* spp. has been called *Herella, Mima, Achromobacter, Alcaligenes, Neisseria, Micrococcus, Diplococcus, Cytophaga,* and *Moraxella* (Bergogne-Bérézin et al., 1987). Ultimately, the oxidase-negative strains of *Moraxella* were named *Acinetobacter* (Baumann et al., 1968).

7.2.2 Epidemiology and Costs of *A. baumannii* Infection

Recently, attention has been directed toward *Acinetobacter* spp., in particular, *A. baumannii* as a result of a large number of *Acinetobacter* infections amongst injured U.S. troops in the Middle East (Scott et al., 2004). Community-onset *Acinetobacter*-pneumonia is a serious disease with high mortality, unless promptly diagnosed and treated (Davis et al., 2014). For *A. baumannii* bacteremia, mortality is high; 54% in one study (Gulen et al., 2015). In addition, *Acinetobacter* spp. has been shown responsible for community-acquired severe, bacteremic pneumonia in which 80% of patients required hospitalization in an intensive care unit (Davis et al., 2014).

In a study of 41 patients with bloodstream (bacteremia) *A. baumannii* infections, two risk factors stood out: prior antibiotic therapy and the presence of a catheter (Gulen et al., 2015). In that study, the cost of hospitalization was $35,000 and the cost of antibiotics was $1,100 (Gulen et al., 2015).

7.2.3 Risk Factors for *A. baumannii* Infection

Risk factors for multi-drug resistant *Acinetobacter* spp. infections include diabetes, chronic pulmonary disease, renal insufficiency, malignancy, prior surgery, employment of a central venous catheter, and antibiotic use (Kanafani et al., 2018). In another study, renal failure, prior antibiotic use, and length of stay prior to infection with multidrug-resistant *Acinetobacter* were all associated with healthcare-associated infection caused by antibiotic-resistant *A. baumannii* (Ellis et al., 2015). Length of hospitalization appears as a major factor in the acquisition of multidrug-resistant *Acinetobacter* spp. infection (Sunenshine et al., 2007), which is consistent with the significant association of infection by multidrug-resistant *Acinetobacter* and residence in rooms with *Acinetobacter* (Munoz-Price et al., 2013).

7.2.4 Sources of *A. baumannii*

Acinetobacter spp. has been recovered from both community and hospital environments. Community environments include natural water and soils (Baumann, 1968), untreated rural drinking water wells (Bifulco et al., 1989), treated water and distribution systems (Narciso-da-Rocha et al., 2013) water, and premise plumbing (Umezawa et al., 2015). "Holy water" has also been shown to contain *A. baumannii* (Rees and Allen, 1996), and *Acinetobacter* spp. has been isolated in high numbers from kitchen sponges (Cardinale et al., 2017). *Acinetobacter* spp. has been isolated from a variety of hospital environments including faucet aerators (Lv et al., 2019), cold-air humidifiers, tap water and washbasins, water baths, ventilator equipment, respirometers, catheters, and liquid hand-lotions and skin cleansing- and disinfecting-solutions (Bergogne-Bérézin et al., 1987). As would be expected from an inhabitant of drinking water systems and premise plumbing, *A. baumannii* is resistant to chlorine (Karumathil et al., 2014). Like other opportunistic waterborne premise plumbing pathogens, *A. baumannii* readily adheres to surfaces and forms biofilms (Qi et al., 2016). *A. baumannii* is an amoeba-resistant microorganism (ARM), with cell numbers increased approximately 10^5-fold in amoeba *A. baumannii* co-cultures compared to number increases of amoebae-free *A. baumannii* cultures (Cateau et al., 2011). Although co-cultures do not accurately capture the growth of intracellular bacteria due to the fact that phagocytosis continues through the co-cultivation, the dramatically increased number of *A. baumannii* cells (Cateua et al., 2011) certainly indicates that *A. baumannii* is an ARM and its numbers are increased in the presence of phagocytic amoebae. Thus, this opportunistic premise plumbing pathogen presents challenges in the treatment of infections. In addition to isolation from terrestrial habitats, *A. baumannii* was isolated from contingency water containers aboard the International Space Station (ISS) (Castro et al., 2004). It is likely that their isolation from water aboard the ISS was due to the colonization of either the water or the containers on the earth before launch.

One of the questions relevant to the identification of sources is the method chosen to fingerprint strains. Common DNA-based fingerprinting techniques include pulsed-field gel electrophoresis (PFGE), multilocus sequence typing (MLST), and whole-genome sequencing (WGS). An investigation of carbapenemase-producing *A. baumannii* showed that the results of PFGE were not the same as obtained by WGS (Bogaty et al., 2018). Likewise in a study of successive *A. baumannii* isolates from a single patient, PFGE- and MLST-results did not agree (Choi et al., 2017). Thus, at present, there is no clear choice of fingerprinting techniques. I would suggest that any fingerprint-comparison study of any of the OPPPs would start successive isolates from a single patient (Choi et al., 2017), but employ all three methods.

7.2.5 Transmission of *A. baumannii*

Like *M. avium* and *M. abscessus*, the aerosol transmission of *Acinetobacter* spp. should be considered. An early article describing *Acinetobacter anitratus* proposed that a hospital outbreak involving 36 patients in different wards was associated with the epidemic strain in the air of patients (Allen and Green, 1987). The aerosol transmission was proposed in the absence of recovery of *A. anitratus* from respiratory equipment (Allen and Green, 1987). The possible airborne transmission of *Acinetobacter* spp. was confirmed in two other independent reports. An *Acinetobacter* spp. outbreak of bloodstream infections in a nursery reported that it was recovered from settle plates in the nursery and nowhere else in other hospital areas

(McDonald et al., 1998). Further, the frequency of *Acinetobacter* spp. bloodstream infections was the highest in summer and associated with increased absolute humidity (McDonald et al., 1998). Likewise, *Acinetobacter* spp. was more frequently recovered from settle plates in an intensive care unit than from surgical wards in the same hospital in Hong Kong (Houang et al., 2001). Recently, the air in a public teaching hospital in Miami, Florida was cultured by settle plate and *A. baumannii* strains isolated whose clonality with patient strains was demonstrated by PFGE and MLST (Munoz-Price et al., 2013) or repetitive extragenic palindromic PCR (*rep*-PCR) (Shimose et al., 2015).

An interesting instance of aerosol transmission was reported, due to the lavage with pulsating water jets (Maragakis et al., 2004). That investigation led to a change in the pulsatile lavage procedure in which disposable suction canisters, which had been previously discarded after use with each patient, were only changed daily or when full (Maragakis et al., 2004). Not only were patients infected with a multidrug-resistant strain of *A. baumannii*, but the same strain (by PFGE) was isolated from the pulsatile lavage device, suction canister, and sink top (Maragakis et al., 2004). The aerosol transmission and isolation from the pulsatile lavage device and from surfaces attest to the environmental persistence of this OPPP.

The persistence of *A. baumannii* in patient rooms is significantly associated with patient infection. In a study of patient rooms, the presence of an *A. baumannii*-infected patient was significantly associated with the presence of *A. baumannii* in the room (Odds ratio of 5.7; Munoz-Price et al., 2013). Only 10% of patient rooms were colonized with *A. baumannii*, in contrast to 39% of rooms colonized with *A. baumannii* for infected patients (Munoz-Price et al., 2013).

7.2.6 Physiological Ecology of *A. baumannii*

A. baumannii has been isolated from drinking water (Bifulco et al., 1989), survives and grows in *Acanthamoeba* sp. (Cateau et al., 2011), and it is approximately 1,000-time more resistant to chlorine than *E. coli* (Karumathil et al., 2014). It is quite likely that the ability of *A. baumannii* to form biofilms (Qi et al., 2016) directly contributes to its tolerance of disinfectants. Evidently, *A. baumannii* presence is associated with higher seasonal temperatures as the frequency of bloodstream infections is the highest during seasons of water temperature (Eber et al., 2011). What is interesting in that temperature-linked increase was that they occurred amongst hospitalized patients (Eber et al., 2011), implicating the impact of climate changes on *A. baumannii* ecology.

In common with *P. aeruginosa, A. baumannii* and other *Acinetobacter* species are killed upon exposure to coupons of copper and copper-metal combinations (e.g., Cu-Zn, Cu-Sn, Cu-Ni) compared to exposure to stainless steel (Różańska et al., 2018). Copper-silver ionization was shown to be able to kill a substantial proportion of planktonic or biofilm-associated cells of *P. aeruginosa, S. maltophilia*, and *A. baumannii* (Shih and Lin, 2010). In spite of 99.99% killing, there was regrowth (i.e., 2- to 3-log increase in number) of *P. aeruginosa* cells after 24 h. No substantial regrowth of either *S. maltophilia* or *A. baumannii* was noted (Shih and Lin, 2010).

7.2.7 Genome Reorganization of *A. baumannii*

Genome reorganization may be a hallmark of OPPPs. It is adaptive in the sense that the generation of phenotypic variants allows the survival of the population in spite of unexpected stresses. Some stresses can be expected by an environmental microorganism, such as exposure to UV-irradiation, desiccation, and toxic oxygen radicals. However, others may not have been encountered in the prior evolutionary history, such as quaternary ammonium compounds. Thus, there is a selection for a toolkit for genetic rearrangement. One example of the results of a toolkit device for genome reorganization is evidence for the role of insertion sequences affecting genomic variability in *A. baumannii* (Gaiarsa et al., 2019). The role of insertion sequence elements in generating high genome variability was shown by a stable persisting clone found in Italian hospitals. That persistent clone of *A. baumannii* (i.e., sequence type 78) has one insertion sequence, namely IS66, intercalated into the *comEC/rec2* gene involved in the acquisition of DNA (Gaiarsa et al., 2019), thus reducing that clone's ability to acquire foreign DNA.

References

Allen KD, Green HT. 1987. Hospital outbreak of multi-resistant *Acinetobacter anitratus*: An airborne mode of spread. *J Hosp Infect.* 9: 110–119.

Baumann P. 1968. Isolation of *Acinetobacter* from soil and water. *J Bacteriol.* 96: 39–42.

Baumann P, Doudoroff M, Stanier RY. 1968. A study of the *Moraxella* group. *J Bacteriol.* 95: 1520–1541.

Bergogne-Bérézin E, Joly-Guillou ML, Vieu JF. 1987. Epidemiology of nosocomial infections due to *Acinetobacter calcoaceticus. J Hosp Infect.* 10: 105–113.

Bifulco JM, Shirley JJ, Bissonnette GK. 1989. Detection of *Acinetobacter* spp. in rural drinking water supplies. *Appl Environ Microbiol.* 55: 2214–2219.

Bogarty C, Mataseje L, Gray A, Lefebvre B, Lévesque S, Mulvey M, Longtin Y. 2018. Investigation of a carbapenemase-producing *Acinetobacter baumannii* outbreak using whole genome sequencing versus a standard epidemiologic investigation. *Antimicrob Resist Infect Contrl.* 7: 140.

Cardinale M, Kaiser D, Lueders T, Schnell S, Egert M. 2017. Microbiome analysis and confocal microscopy of used kitchen sponges reveal massive colonization by *Acinetobacter, Moraxella* and *Chryseobacterium* species. *Sci Rpts.* 7: 5791.

Castro VA, Thrasher AN, Healy M, Ott CM, Pierson DL. 2004. Microbial characterization during the early habitation of the International Space Station. *Microb Ecol.* 47: 119–126.

Cateau E, Verdon J, Fernandez B, Hechard Y, Rodier M-H. 2011. *Acanthamoeba* sp. promotes the survival and growth of *Acinetobacter baumannii. FEMS Microbiol Lttr.* 319: 19–25.

Choi HJ, Kil MC, Choi J-Y, Kim SJ, Park K-S, Kim Y-J, Ko KS. 2017. Characterisation of successive *Acinetobacter baumannii* isolates from a deceased haemophagocytic lymphohistiocytosis patient. *Int J Antimicrob Agents*. 49: 102–106.

Davis JS, McMillan M, Swaminathan A, Kelly JA, Piera KE, Baird RW, Currie BJ, Anstey NM. 2014. A 16-year prospective study of community-onset bacteremic *Acinetobacter* pneumonia. *Chest*. 146: 1038–1045.

Dijkshoorn L, Nemec A, Seifert H. 2007. An increasing threat in hospitals: Multidrug-resistant *Acinetobacter baumannii*. *Nat Revs*. 5: 939–951.

Eben MR, Shardell M, Schweizer ML, Laxminarayan R, Perencevich EN. 2011. Seasonal and temperature-associated increases in Gram-negative bacterial bloodstream infections among hospitalized patients. *PLoS ONE*. 6: e25298.

Ellis D, Cohen B, Liu J, Larson E. 2015. Risk-factors for hospital-acquired antimicrobial-resistant infection caused by *Acinetobacter baumannii*. 4: 40.

Gaiarsa S, Bitar I, Cpmandatore F, Corbella M, Piazza A, Scaltriti E, Villa L, Postiglione U, Marone P, Nucleo E, Pongolini S, Migliavacca R, Sassera D. 2019. Can insterion sequences proliferation influence genomic plasticity? Comparative analysis of *Acinetobacter baumannii* sequence type 78, a persistent clone in Italian hospitals. *Front Microbiol*. 10: 2080.

Gallagher LA, Ramage E, Weiss EJ, Radey M, Hayden HS, Held KG, Huse HK, Zurawaski DV, Brittnacher MJ, Manoil C. 2015. Resources for genetic and genomic analysis of emerging pathogen *Acinetobacter baumannii*. *J. Bacteriol*. 197: 2027–2035.

Gulen TA, Guner R, Celikbilek N, Keske S, Tasyaran M. 2015. Clinical importance and cost of bacteremia caused by nosocomial multi drug resistant *Acinetobacter baumannii*. *Int. J. Infect. Dis*. 38: 32–35.

Houang ETS, Chu YW, Leung CM, Chu KY, Berlau J, Ng KC, Cheng AFB. 2001. Epidemiology and infection control implications of *Acinetobacter* spp. in Hong Kong. *J Clin Microbiol*. 39: 228–234.

Kanafani ZA, Aehreddine N, Tayyar R, Sfeir J, Araj GF, Matar GM, Kanj SS. 2018. Multi-drug resistant *Acinetobacter* species: A seven-year experience from a tertiary care center in Lebanon. *Antimicrob Resist Infect Contrl*. 7:9.

Karumathil DP, Yin H-B, Kollanoor-Johny A, Venkitanarayanan K. 2014. Effect of chlorine exposure on the survival and antibiotic gene expression of multidrug resistant *Acinetobacter baumannii. Int J Environ Res Publ Hlth.* 11: 1844–1854.

Logan LK, Gandra S, Trett A, Weinstein RA, Laxminarayan R. 2018. *Acinetobacter baumannii* resistance trends in children in the United States, 1999–2012. *J Ped Infect Dis.* 8: 136–142.

Lv Y, Xiang Q, Ying ZJ, Ying F, Wu YJ, Zeng B, Yu H, Cai HM, Wei QD, Wang C, Chen J, Wang H. 2019. Faucet aerators as a reservoir for Carbapenem-resistant *Acinetobacter baumannii*: A healthcare-associated infection outbreak in a neurosurgical intensive care unit. *Antimicrob Rest Infect Cntrl.* 8: 205.

Maragakis LL, Cosgrove SE, Song X, Kim D, Rosenbaum P, Ciesla N, Srinivasan A, Ross T, Carroll K, Perl TM. 2004. An outbreak of multidrug-resistant *Acinetobacter baumannii* associates with pulsatile lavage wound treatment. *J Am Med Assoc.* 292: 3006–3011.

McDonald LC, Walker M, Carson L, Arduino M, Aguero S, Gomez P, McNeil P, Jarvis WR. 1998. Outbreak of *Acinetobacter* spp. bloodstream infections in a nursery associated with contaminated aerosols and air conditioners. *Ped Infect Dis J.* 17: 716–722.

Munoz-Price LS, Fajardo-Aquino Y, Arheart KL, Cleary T, DePascale D, Pizano L, Namias N, Rivera JI, O'Hara JA, Doi Y. 2013. Aerosolization of *Acinetobacter baumannii* in a trauma ICU. *Crit Care Med.* 41: 1915–1918.

Munoz-Price LS, Namian N, Cleary T, Fajardo-Aquino Y, DePascale D, Arheart KL, Rivera JI, Doi Y. 2013. *Acinetobacter baumannii*: Association between environmental contamination of patient rooms and occupant status. *Infect Contr Hosp Eidemiol.* 34: 517–520.

Munoz-Price LS, Weinstein RA. 2008. *Acinetobacter* infection. *N Engl J Med.* 359: 1271–1281.

Narciso-da-Rocha C, Vaz-Moreira I, Svensson-Stadler L, Moore ERB, Manaia CM. 2013. Diversity and antibiotic resistance of *Acinetobacter* spp. in water from the source to the tap. *Appl Microbiol Biotechnol.* 97: 329–340.

Qi L, Li H, Zhang C, Liang B, Li J, Wang L, Du X, Liu X, Qiu S, Song H. 2016. Relationship between antibiotic resistance, biofilm formation, and biofilm-specific resistance in *Acinetobacter baumannii. Front Microbiol.* 7: 483.

Rees JC, Allen KD. 1996. Holy water: A risk factor for hospital-acquired infection. *J Hosp Infect.* 32: 51–55.

Roca I, Espinal P, Vila-Farrés X, VilaJ. 2012. The *Acinetobacter baumannii* oxymoron: Commensal hospital dweller turned pan-drug-resistant menace. *Front Microbiol.* 3:148. doi: 10.3389/fmicb.2012.00148.

Różańska A, Chmielarczyk A, Romaniszyn D, Majka G, Bulanda M. 2018. Antimicrobial effect of copper alloys on *Acinetobacter* species isolated from infections and hospital environment. *Antimicrob Resist Infect Contrl.* 7: 10.

Scott PT, Peterson K, Fishbain J, Craft DW, Ewell AJ, Moran K, Hack DC, Deye GA, Riddell S, Christopher G, Mancuso JD, Petruccelli BP, Endy T, Lindler L, Davis K, Milstrey EG, Brosch L, Pol J, Blankenship CL, Witt CJ, Malone JL, Tornberg DN, Srinivasan A. 2004. *Acinetobacter baumanii* infections among patients at military medical facilities treating injured U.S. service members, 2002–2004. *Morbid Mortal Week Rpt.* 53: 1063–1066.

Shih H-S, Lin YE. 2010. Efficacy of copper-silver ionization in controlling biofilm- and plankton-associated waterborne pathogens. *Appl Environ Microbiol.* 76: 2032–2035.

Shimose LA, Doi Y, Bonomo RA, De Pascale D, Viau RA, Cleary T, Namias N, Kett DH, Munoz-Price LS. 2015. Contamination of ambient air with *Acinetobacter baumannii* on consecutive inpatient days. *J Clin Microbiol.* 53: 2346–2348.

Sunenshine RH, Wright M-O, Maragakis LL, Harris AD, Song X, Hebden J, Cosgrove SE, Anderson A, Carnell J, Jernigan DB, Kleinbaum DG, Perl TM, Standiford HC, Srinivasan A. 2007. Multidrug-resistant *Acinetobacter* infection mortality rate and length of hospitalization. *Emerg Infect Dis.* 13: 97–103.

Unezawa K, Asai S, Ohshima T, Iwashita H, Ohashi M, Sasaki M, Kaneko A, Inokuchi S, Miyachi H. 2015. Outbreak of drug-resistant *Acinetobacter baumannii* ST219 caused by oral care using tap water from contaminated hand hygiene sinks as reservoir. *Am J Infect Contrl.* 43: 1249–1251.

7.3 *Sphingomonas paucimobilis*

7.3.1 Introduction to *Sphingomonas* spp.

Members of the genus *Sphingomonas* are both opportunistic pathogens and important environmental contributors to water and soil. Major distinguishing features include the presence of sphingolipids and the lack of 3-hydroxy fatty acid that is normally a major component of Gram-negative lipopolysaccharides (Kawahara et al., 1990). Several sphingomonads, specifically *Sphingomonas macrogoltabidus, Sphingomonas sanguis,* and *Sphingomonas terrae,* utilize polyethylene glycol (e.g., PEG 4,000) for growth (Takeuchi et al., 1993).

The most commonly reported opportunistic pathogen of the genus is the deep-yellow-pigmented, aerobic *S. paucimobilis,* earlier identified as *Pseudomonas paucimobilis* (Yabuuchi et al., 1990). The yellow-pigmented colonies can be misidentified as *Flavobacterium* sp. *S. paucimobilis* is distinguished by slow motility and has been reported to cause meningitis, septisemia, bacteremia, peritonitis, and wound infections. The novel lipid A-type glycolipid has been reported as expressing (Smalley, 1982) or lacking endotoxin activity (Kanahara et al., 1980). *S. paucimobilis* has been recovered from infusion fluids, tube-rinsing water, irrigation fluids, and non-sterile water in operating rooms (Yabuuchi et al., 1990).

7.3.2 *S. paucimobilis* Infections and Risk Factors

S. paucimobilis-bacteremia was reported in 16 individuals over the April 2004–April 2008 period from a Taiwanese hospital (Lin et al., 2010). Common co-morbidities included malignancy, immunosuppression, and diabetes and both primary bacteremia and catheter-related bloodstream infections were encountered within the studied population (Lin et al., 2010). Both severe and invasive infections (e.g., septic arthritis or osteomyelitis) have been reported (Ryan and Adley, 2010). Based on the report of *S. paucimobilis*-meningitis and -ventriculitis in a patient undergoing immunosuppression therapy (Bolen et al.,

2015), reduced immune competence appears to be a risk factor for *S. paucimobilis* infection. Recurrent *S. paucimobilis* bacteremia was linked to the colonization of the hospital water system by that OPPP (Perola et al., 2002). *S. paucimobilis* was also shown to cause a pacemaker pocket infection in a child with a hemangioma (Shi et al., 2016) and bone and soft tissue infections (El Beaino et al., 2018). Another *Sphingomonas* species, *S. koreensis*, responsible for infections for approximately 10 years, was present in sinks of patient rooms in the NIH Clinical Center, Bethesda, MD (Johnson et al., 2018). Analysis of archived *S. paucimobilis* strains from the hospital and patients demonstrated intermittent recovery of a single clone, demonstrating the persistence of a single type over a 10-year period (Johnson et al., 2018).

7.3.3 Epidemiology and *S. paucimobilis* Infection Sources

Both community- and hospital-acquired *S. paucimobilis* infections have been reported. In a study of *S. paucimobilis* infections in a medical center in Taiwan, 29 of 55 (53%) *S. paucimobilis*-infected patients had community-acquired infections and 13 of 29 (45%) had bacteremia (Toh et al., 2011). Within that group, diabetes and alcoholism were significant risk factors for *S. paucimobilis* bacteremia (Toh et al., 2011). Interestingly, the most common presentation of hospital-acquired *S. paucimobilis* infection was pneumonia (Toh et al., 2011). Another set of risk factors, namely chronic obstructive pulmonary disease (COPD), chronic asthmatic bronchitis, and a 25-year history of exposure to cotton fibers, were linked to the acquisition of *S. paucimobilis* bacteremia (Morrison and Shulman, 1986). One interesting feature of that isolate was its growth at 25 °C, but not at 42 °C (Morrison and Shulman, 1986). Another community-acquired infection was in a diabetic patient with *S. paucimobilis* bacteremia and septic pulmonary emboli (Kuo et al., 2009).

Infection sources for *S. paucimobilis* hospital-acquired infections have been traced to "purified" hospital water (Perez del Molino and Garcia-Ramos, 1989), contaminated solutions (e.g., distilled water hemodialysis fluid), non-sterile drug solutions (Ryan and Adley, 2010), hemodialysis fluid (Phillips et al., 1991), *S. paucimobilis*-contaminated catheters (Lanoix et al., 2012), hospital hot water lines (Crane et al., 1981), nebulizer fluid (Oie et al., 2006), ventilator temperature probes (Lemaitre et al., 1996), and oxygen flow meters in an intensive care unit (Meric et al., 2009). The common thread in all those instances cited is the assumption that distilled, purified, or even tap water is sterile. Clearly, it is not sterile. The isolates of *S. paucimobilis* from the flow meters shared the same pulsed-field gel electrophoresis (PFGE) band pattern as those of the patients (Meric et al., 2009). Evidently, *S. paucimobilis* is a member of the endemic hospital flora (Morrison and Shulman, 1986). The presence of *S. paucimobilis* in a hot water line in an intensive care unit was associated with an outbreak of *S. paucimobilis* colonization and infection among patients receiving ventilation assistance (Crane et al., 1981). Bottles filled with hot water used to rinse tracheal suction tubing were found to contain *S. paucimobilis* (Crane et al., 1981). Unfortunately, the hospital was unable to eradicate *S. paucimobilis* from the hot water line (Crane et al., 1981), suggesting that the microorganism is relatively disinfectant-resistant, possibly due to its ability to adhere and form biofilms in pipes. In another instance of hospital-acquired *S. paucimobilis* infection, ventilators used for neonates were shown to be colonized (Lemaitre et al., 1996). The *S. paucimobilis* strain in the ventilators was traced to the use of a non-sterile temperature probe (Lemaitre et al., 1996). All isolates were shown to be clonal by arbitrarily primed polymerase chain reaction (AP-PCR) fingerprinting (Laemaitre et al., 1996). Nosocomial catheter-associated *S. paucimobilis* infections have also been reported (Hsueh et al., 1998). Interestingly, in two patients with catheter-associated infections, multiple isolates of *S. paucimobilis* were isolated from the patients with a variety of different AP-PCR fingerprints were obtained (Hsueh et al., 1998). *S. paucimobilis* isolates were recovered from a dialysate, humidifier liquid, and from tap water, all displayed unique AP-PCR fingerprints, none of which matched those from the patients (Hsueh et al., 1998). As noted above, *S. paucimobilis* is a normal hospital tap water resident. That was underscored by the report of two recurrent *S. paucimobilis*-bacteremia infections amongst neutropenic patients over a period of two years (Perola et al., 2002). One strain of *S. paucimobilis* isolated from tap water in the hematological unit had the same AP-PCR fingerprint as that from one patient; however, the hospital harbored a variety of *S. paucimobilis* strains based on AP-PCR fingerprints (Perola et al., 2002).

A pseudo-epidemic of *S. paucimobilis* was reported, due to contaminated saline irrigation fluid used for sinus irrigation (Faden et al., 1981). The source of contamination was shown to be due to leaks from a sink, under which the sterile saline bottles were stored. A survey of 26 sinks in the hospital demonstrated that 16 (62%) were colonized by *S. paucimobilis* with numbers from 20–120,000 colonies per liter (Faden et al., 1981). The highest *S. paucimobilis* numbers were from sinks with faucet aerators prompting the removal of aerators or their frequent replacement if removal was impossible (Faden et al., 1981).

7.3.4 Unique *S. paucimobilis* Features

There are three unique features of *S. paucimobilis*: (a) production of the deep yellow pigment in colonies, (b) slow motility of cells (paucimobilis), and (c) presence of a sphingolipid with the long-chain base dihydrosphingosin, ubiquinone-10, and 2-hydroxymyristic acid (2-OH C14:0) (Ryan and Adley, 2010). Depending upon the primary isolation agar medium, *S. paucimobilis* colonies can be deeply yellow-colored; whereas, on other media (e.g., R2A agar used for water analysis), the pigmentation is reduced. Slow motility means that motile cells move across the microscope field slowly compared to cells of *Pseudomonas aeruginosa*. The frequency of motile cells is the same whether one is examining a suspension of *S. paucimobilis* or *P. aeruginosa*; not all cells are motile at any one time, just a fraction. Unfortunately, there is no direct, simple test for sphingolipids that would provide rapid, unambiguous confirmation of identity.

As noted above, there is conflict in the literature concerning the endotoxin activity of the lipooligosaccharide of *S. paucimobilis.*

Kawahara et al. (1990) reported an absence of endotoxic activity in the *Limulis* test. In contrast, Smalley (1982) reported that 7 of 7 strains of S. *paucimobilis* demonstrated endotoxin activity in the *Limulus* amoebocyte lysate test. In the absence of details of the growth and cell fraction (e.g., whole cells or purified cellular subfraction) tested, a definitive answer cannot be provided at present.

7.3.5 Ecology of *S. paucimobilis*

S. paucimobilis is widely distributed in natural soil and water and in human-engineered water distribution systems. As reviewed by Ryan and Adley (2010), *S. paucimobilis* has been isolated from a variety of natural water habitats, including sea water, river water, waste water, and mineral water. Further, sphingomonads are common in marine environments (Cavicchioli et al., 1999) and have been associated with diseased corals (Richardson et al., 1998). In hospitals, *S. paucimobilis* has been isolated from drinking water distribution systems, and water-containing equipment used in hospitals, such as therapy baths (Ryan and Adley, 2011) and heater-coolers (Chand et al., 2017). Sadly, a recent report documented *S. paucimobilis* bacteremias and the presence of *S. paucimobilis* in tap water at the Roswell Park Comprehensive Cancer Center, due to the diversion of hydromorphone in vials for patient treatment and its replacement with tap water (Wasira et al., 2019). Sphingomonads have even been isolated from cleanrooms used for the assembly of spacecraft and testing components (La Duc et al., 2007) and *S. paucimobilis* has been recovered from water samples aboard the International Space Station (Castro et al., 2004).

As noted above, the frequent colonization of hospital sink water with *S. paucimobilis* in numbers as high as 120,000 colonies per liter (Faden et al., 1981) underscores the comment that *S. paucimobilis* is a member of the endemic hospital flora. (Morrison and Shulman, 1986). Further, *S. paucimobilis* can persist for long periods of time in hospital water systems; up to 2 years for a particular clone (Perola et al., 2002). Evidence that one hospital was unable to eradicate *S. paucimobilis* from the hot water lines

(Crane et al., 1981), suggests that the microorganism is relatively disinfectant-resistant, possibly due to its ability to adhere and form biofilms in pipes.

One strategy for the reduction of *S. paucimobilis* numbers in hospital water lines apparently is to remove tap aerators, as the highest *S. paucimobilis* numbers were from sinks with faucet aerators (Faden et al., 1981). Where aerators cannot be removed, frequent replacement might work (Faden et al., 1981). Another approach to reducing *S. paucimobilis* numbers was suggested by an article demonstrating its high susceptibility to heavy metals (Tada and Inoue, 2000). Growing under oligotrophic conditions (e.g., drinking water), the EC_{50} concentrations of Zn^{+2} and Cu^{+2} were 10^{-4} mM (Tada and Inoue, 2000), the concentrations that could be reached in systems using galvanized or copper pipes. However, that would not work for *M. avium,* as it is quite resistant to those metals; in fact, copper pipe biofilms in a pilot distribution system were almost exclusively composed of *M. avium* (Norton et al., 2004).

References

Bolen RD, Palavecino E, Gomadam A, Balakrishnan N, Datar S. 2015. *Sphingomonas paucimobilis* meningitis and ventriculitis in an immunocompromised host. *J Neurol Sci.* 359: 18–20.

Castro VA, Thrasher AN, Healy M, Ott CM, Pierson DL. 2004. Microbial characterization during the early habitation of the International Space Station. *Microb Ecol.* 47: 119–126.

Cavicchioli R, Fegatella F, Ostrowski M, Eguchi M, Gottschal J. 1999. Sphingomonads from marine environments. *J Indust Microbiol Biotechnol.* 23: 268–272.

Chand M, Lamagni T, Kranzer K, Hedge J, Moore G, Parks S, Collins S, del Ojo Elias C, et al. 2017. Insidious risk of severe *Mycobacterium chimera* infection in cardiac surgery patients. *Clin Infect Dis.* 64: 335–342.

Chang HS, Wi YM, Moon SY, Kang C-I, Son JS, Ko KS, Chung DR, Lee NY, Song J-H, Peck KR. 2008. Clinical features of treatment outcomes of infections caused by *Sphingomonas paucimobilis. Infect Contr Hosp Infect.* 29: 990–992.

Cheong HS, Wi YM, Moon SY, Kang C-I, Son JS, Ko KS, Chung DR, Lee NY, Song J-H, Peck KR. 2008. Clinical features and treatment outcomes of infections caused by Sphingomonas paucimobilis. *Infect Contr Hosp Epi.* 29: 990–992.

Crane LR, Tagle LC, Palutke WA. 1981. Outbreak of *Pseudomonas paucimobilis* in an intensive care facility. *J Am Med Assoc.* 246: 985–987.

El Beaino M, Fares J, Malek A, Hachem R. 2018. *Sphingomonas paucimobilis*-related bone and soft-tissue infections: A systematic review. *Intl J Infect Dis.* 77: 68–73.

Faden H, Britt M, Epstein B. 1981. Sinus contamination with *Pseudomonas paucimobilis*: a pseudoepidemic due to contaminated irrigation fluid. *Infect Contr.* 2: 233–235.

Holmes B, Owen RJ, Evans A, Malnick H, Willcox WR. 1977. *Pseudomonas paucimobilis*, a new species isolated from human clinical specimen, the hospital environment, and other sources. *Int J System Bacteriol.* 27: 133–146.

Hsueh P-R, Teng L-J, Yang P-C, Chen Y-C, Pan H-J, Ho S-W, Luh K-T. 1998. Nosocomial infections caused by *Sphingomonas paucimobilis*: Clinical features and microbiological characteristics. *Clin Infect Dis.* 26: 676–681.

Johnson RC, Deming C, Conlan S, Zellmer CJ, Park M, Weingarten RA, Less J, Dekker JP, Frank KM, Musser KA, McQuiston JR, Henderson DK, Lau AF, Palmore TN, Segre JA. 2018. Investigation of a cluster of *Sphingomonas koreensis* infections. *N Engl J Med.* 379: 2529–2539.

Kawahara K, Matsuura M, Danbara H. 1990. Chemical structure and biological activity of lipooligosaccharide isolated from *Sphingomonas paucimobilis*, a Gram-negative bacterium lacking usual lipopolysaccharide. *Jpn J Med Sci Biol.* 43: 250.

Kuo I-C, Lu P-L, Lin W-R, Lin C-Y, Chang Y-W, Chen T-C, Chen Y-H. 2009. Sphingomonas paucimobilis bacteraemia and septic arthritis in a diabetic patient presenting with septic pulmonary emboli. *J Med Microbiol.* 58: 1259–1263.

La Duc MT, Dekas A, Osman S, Moissl C, Newcomb D, Venkateswaran K. 2007. Isolation and characterization of bacteria capable of tolerating the extreme conditions of clean room environment. *Appl Environ Microbiol.* 73: 2600–2611.

Lanoix J-P, Hamdad F, Borel A, Thomas D, Salle V, Smail A, El Samad Y, Schmit J-L. 2012. *Sphingomonas paucimobilis* bacteremia related to intra-venous human immunoglobulin injections. *Med infect Malad.* 42: 36–43.

Lemaitre D, Elaichouni A, Hundhausen M, Claeys G, Vanhaesebrouck P, Vaneechoutte M, Verschraegen G. 1996. Tracheal colonization with *Sphingomonas paucimobilis* in mechanically ventilated neonates due to contaminated ventilator probes. *J Hosp Infect.* 32: 199–206.

Lin J-N, Lai C-H, Chen Y-H, Lin H-L, Huang C-K, Chen W-F, Wang J-L, Chung H-C, Liang S-H, Lin H-H. 2010. *Sphingomonas paucimobilis* bacteremia in humans: 16 case reports and a literature review. *J Microbiol Immunol Infect.* 43: 35–42.

Meric M, Willke A, Kolayli F, Yavuz S, Vahaboglu H. 2009. Water-borne *Sphingomonas paucimobilis* epidemic in an intensive care unit. *J Infect.* 58: 253–255.

Morrison AJ Jr, Shulman JA. 1986. Community-acquired bloodstream infection caused by Pseudomonas paucimobilis: Case report and review of the literature. *J Clin Microbiol.* 24: 853–855.

Norton CD, LeChevallier MW, Falkinham JO III. 2004. Survival of *Mycobacterium avium* in a model distribution system. *Water Res.* 38: 1457–1466.

Novikova N, De Boever P, Poddubko S, Deshevaya E, Polikarpov N, Rakova N, Coninx I, Mergeay M. 2006. Survey of environmental biocontamination on board the International Space Station. *Res Microbiol.* 157: 5–12.

Oie S, Makieda D, Ishida S, Okano Y, Kamiya A. 2006. Microbial contamination of nebulization solution and its measures. *Biol Pharm Bull.* 29: 503–507.

Perez del Molino ML, Garcia-Ramos R. 1989. *Pseudomonas paucimobilis* in purified water for hospital use. *J Hosp Infect.* 14: 373–374.

Perola O, Nousiainen T, Suomalainen S, Aukee S, Kärkkäinen U-M, Kauppinen J, Ojanen T, Katila M-L. 2002. Recurrent *Sphingomonas paucimobilis*-bacteraemia associated with a multi-bacterial water-borne epidemic among neutropenic patients. *J Hosp Infect.* 50: 196–201.

Phillips G, Fleming LW, Stewart WK, Hudson S. *Pseudomonas paucimobilis* contamination in haemodialysis fluid. *J Hosp Infect.* 17: 70–71.

Richardson LL, Goldberg WM, Kuta KG, Aronson RB, Smith GW, Ritchie KB, Halas JC, Feingold JS, Miller SL. Florida's mystery coral-killer identified. *Nature*. 392: 557–558.

Ryan MP, Adley CC. 2010. *Sphingomonas paucimobilis*: A persistent Gram-negative nosocomial infectious organism. *J Hosp Infect*. 75: 153–157.

Shi X, Liu R, 2016. *Sphingomonas paucimobilis* causing pacemaker pocket infection in a pediatric patient with a hemangioma. *Am J Infect Contr*. 44: 617–618.

Smalley DL. 1982. Endotoxin-like activity in *Pseudomonas paucimobilis* (group IIk biotype 1) and *Flavobacterium multivorum* (group IIk biotype 2). *Experientia*. 38: 1483–1484.

Tada Y, Inoue T. 2000. Use of oligotrophic bacteria for the biological monitoring of heavy metals. *J Appl Microbiol*. 88: 154–160.

Takeuchi M, Kawai F, Shimada Y, Yokota A. 1993. Taxonomic study of polyethylene-utilizing bacteria: Emended description of the genus *Sphingomonas* and new descriptions of *Sphingomonas macrogoltabidus* sp. nov., *Sphingomonas sanguis* sp. nov. and *Sphingomonas terrae* sp. nov. *System Appl Microbiol*. 16: 227–238.

Toh H-S, Tay H-T, Kuar W-K, Weng T-C, Tang H-J, Tan C-K. 2011. Risk factors associated with *Sphingomonas paucimobilis* infection. *J Microbiol Immunol Infect*. 44: 289–295.

Wasiura J, Segal BH, Mullin KM. 2019. Cluster of *Sphingomonas paucimobilis* bacteremias linked to diversion of intravenous hydromorphone. *N Engl J Med*. 381: 584–585.

Yabuuchi E, Yano I, Oyaizu H, Hashimoto Y, Ezaki T, Yamamoto H. 1990. Proposals of *Sphingomonas paucimobilis* gen. nov. and comb nov. *Sphingomonas parapaucimobilis* sp. nov. *Sphingomonas yanoikuyae* sp. nov., *Sphingomonas adhaesiva* sp. nov., *Sphingomonas capsulata* comb. nov., and two genospecies of the genus *Sphingomonas*. *Microbiol Immunol*. 34: 99–119.

7.4 *Methylobacterium* Species

7.4.1 Introduction

I first encountered *Methylobacterium* spp. in a paper describing shower-curtain microbiomes (Kelley et al., 2004) and then later as one of the prevalent showerhead biofilm bacteria, where their presence was associated with the absence of nontuberculous mycobacteria (Feazel et al., 2009). We confirmed the observation that *Methylobacterium* presence indicated the absence of *M. avium* in homes of *M. avium*-infected individuals in Wynnewood, PA (Falkinham et al., 2016). Those observations led us to discover that *Methylobacterium* spp. cells in biofilms were able to inhibit the adherence of *Mycobacterium* spp. cells to plumbing surfaces (Muńoz-Egea et al., 2017). However, as my colleagues and I studied *Methylobacterium*, I discovered that they were opportunistic premise plumbing pathogens (OPPPs).

Methylobacterium spp. are Gram-negative opportunistic pathogens in premise plumbing that form distinctive pink-pigmented colonies on a variety of agar media that were originally classified as *Pseudomonas* or *Protomonas* and members of the α2 Proteobacteria (Green and Bousfield, 1982; Green and Ardley, 2018). In broth culture, *Methylobacterium* spp. form aggregates that are also pink (Gallego et al., 2006). Pigmentation is due to the production of carotenoid pigments. *Methylobacterium* spp. are able to grow on one-carbon compounds such as formate, formaldehyde, methanol, methylamine, and a wide range of multi-carbon compounds as a sole source of carbon and energy (Green, 2006). Further, *Methylobacterium* spp. are capable of fixing atmospheric nitrogen (Sy et al., 2001).

Methylobacterium spp. are plant epiphytes that can stimulate plant growth through their production of phytochromes (Ivanova et al., 2001; Koeing et al., 2002), which help plants to fight pathogens (Holland and Polacco, 1994), and enhance the production of the strawberry furanoid, flavor compounds (Verginer et al., 2010). Recently it has been shown that *Methylobacterium* spp. comprise the highest proportion of bacterial genera in organic-, but not conventionally-produced, apples (Wassermann et al., 2019).

7.4.2 Disease and Risk Factors

Methylobacterium spp. are linked to healthcare-associated infections (Sanders et al., 2000), including those in immunosuppressed individuals (Hornei et al., 1999). Peritoneal dialysis may also be a risk factor for *Methylobacterium* spp. infection (Rutherford et al., 1988). *Methylobacterium* spp. have been isolated from blood, cerebrospinal fluid, bone marrow, and peritoneal fluid (Kovaleva et al., 2014). Catheter-associated infections have also been reported (Kovaleva et al., 2014).

A contaminated endoscope was shown to be the source of *Methylobacterium* spp. bacteremia (Imbert et al., 2005) and pseudo-outbreaks of *Methylobacterium mesophilicum* infection were traced to a contaminated automated endoscope washer (Flournoy et al., 1992; Kressel and Kidd, 2001). Further, activated 2% glutaraldehyde was found to be contaminated with a strain of *Methylobacterium* spp. (Webster et al., 1996), suggesting they are resistant to antimicrobials.

7.4.3 *Methylobacterium* spp. Epidemiology and Infection Sources

Members of the *Methylobacterium* genus are found in a wide variety of natural habitats, including soil, dust, air, freshwater, and aquatic sediments (Hirashi et al., 1995) as well as plants (Ivanova et al., 2001; Koeing et al., 2002; Holland and Polacco, 1994; Wassermann et al., 2019). *Methylobacterium* spp. have been isolated from drinking water systems, hospital tap water, and dental units (Flournoy et al., 1992; Kovalea et al., 2014). They can be easily recognized in homes as they form pink slime in shower surfaces and on shower curtains and form pink rings at the water–air interface in commodes and fountains (Kelley et al., 2005; Yano et al., 2013). *Methylobacterium extorquens* has been isolated from amoeba from drinking water systems, making it a member of the amoeba-resistant microorganism (ARM) group (Thomas et al., 2006). One species, *Methylobacterium tardum*, was isolated from dust samples (Mora et al., 2016) and *Methylobacterium fujisawaense* from a humidity condensate processor, water dispensing unit, and contingency water containers aboard the International Space Station (Castro et al., 2004).

7.4.4 Ecology of *Methylobacterium* spp.

Methylobacterium spp. are common inhabitants, colonists, not contaminants, of drinking water and drinking water distribution systems (Raja et al., 2008) and have been detected by DNA isolation and 16S rRNA gene sequence from showerheads across the United States (Feazel et al., 2009). In fact, *Methylobacterium* spp. were the second most prevalent bacterial group after the *Mycobacterium* in the showerhead samples (Feazel et al., 2009). *Methylobacterium* spp. are able to persist in drinking water due to their chlorine resistance (Hiraishi et al., 1995), surface adherence, biofilm formation, and desiccation tolerance (Kaneko and Hiraishi, 1991; Hugenholtz et al., 1995; Yano et al., 2013).

Methylobacterium-containing biofilms have been found on surfaces in automobile air-conditioning systems (Simmons et al., 1999), shower curtains (Kelley et al., 2004), and showerheads (Feazel et al., 2009). *Methylobacterium* spp. cells in biofilms exhibit resistance to chemical disinfectants. For example, *Methylobacterium* spp. biofilms have been found to survive one-hour exposure to 100 ppm of sodium hypochlorite, a concentration that is 100-times higher than the 1 ppm concentration used for drinking water disinfection (Simoes et al., 2009). This is in keeping with their isolation from water supplies and other chlorinated environments (Furuhata and Koike, 1993). The resistance of *M. extorquens* to free residual chlorine has been found to be moderate; the $CT_{99.9\%}$ value (the product of disinfectant concentration and time to reach 3 logs units of cell death) was 1.5 mg-min/l compared to *E. coli* whose CT value is 0.09 mg-min/l (Furuhata et al., 1989). *Methylobacterium aquaticum* strains isolated from tap water were all found to be mildly resistant to chlorine having average CT values of 0.89 mg-min/l (Furuhata et al., 2011).

Regrowth of *M. extorquens* biofilms occurred following seven days after an exposure to 1% peracetic acid (Kovela, 2013). *M. mesophilicum* biofilms on the tubing in an automated endoscope reprocessor were the source of an *M. mesophilicum* infection outbreak in patients after bronchoscopy (Kressel and Kidd, 2001). Further, the presence of an established *Methylobacterium* spp. biofilm has been found to substantially reduce adherence and biofilm formation by *M. avium* to plumbing surfaces (Muńoz-Egea et al., 2017).

Aggregation of *Methylobacterium* spp. cells appears to be a characteristic, as it has been reported in all species characterized to date, including *Methylobacterium adhaesivum* AR27 (Gallego et al., 2006) and *Methylobacterium funariae* F3.2 (Schauer and Kutschera, 2011). Plant growth-promoting *Methylobacterium* spp. strains have also been found to aggregate in high C/N growth conditions (Joe et al., 2013). A bacterial aggregate formation may be driven by the same forces that result in biofilm formation, namely hydrophobic interactions (Kos et al., 2003).

Methylobacterium spp. are recognized as resistant to elevated (>50 °C) temperatures (Orphan, 1999). *Methylobacterium* spp. have been isolated from hot tap water in a household (Soucie and Schuler, 2006) suggesting that the laboratory-measured resistance (Orphan, 1999) is of ecological consequence. *Methylobacterium radiotolerans* has been isolated from a clean room (La Duc et al., 2007), suggesting its ability to survive in such clean environments and under irradiation. The PhyR protein in *M. extorquens* has been revealed to positively regulate genes involved in the resistance to heat shock and other environmental stressors (Gourion et al., 2007).

7.4.5 *Methylobacterium* spp. as Anti-OPPP Bacteria

I have included the *Methylobacterium* spp. in this volume as they are one of the most common waterborne opportunistic pathogens. They have all the characteristics necessary for persistence and growth in premise plumbing, namely biofilm-formation to prevent washout, disinfectant resistance to permit survival after water treatment process disinfection, and desiccation resistance to permit survival when a drinking water system lacks water and biofilms dry. In addition, it has come to mind that *Methylobacterium* spp. can inhibit *Mycobacterium* spp. surface adherence and biofilm formation (Munoz-Egea et al., 2018), *Methylobacterium* spp. cells or cell fractions could be used to prevent *Mycobacterium* spp. biofilm formation on premise plumbing, showerheads, and medical equipment.

References

Angenent LT, Kelley ST, St Armand A, Pace NR, Hernandez MT. 2005. Molecular identification of potential pathogens in water and air of a hospital therapy pool. *Proc Natl Acad Sci USA*. 102: 4860–4865.

Castro VA, Thrasher AN, Healy M, Ott CM, Pierson DL. 2004. Microbial characterization during the early habitation of the International Space Station. *Microb Ecol.* 47: 119–126.

Collier SA, Stockman LJ, Hicks LA, Garrison LE, Zhou FJ, Beach MJ. 2012. Direct healthcare costs of selected diseases primarily or partially transmitted by water. *Epidemiol. Infect.* 140: 2003–2013.

Cuttelod M, Senn L, Terletskiy V, Nahimana I, Petignat C, Eggimann P, Bille J, Prod'hom G, Zanetti G, Blanc DS. 2011. Molecular epidemiology of *Pseudomonas aeruginosa* in intensive care units over a 10-year period (1998–2007). *Clin Microbiol Infect.* 17: 57–62.

Donlan RM, Costerton JW. 2002. Biofilms: Survival mechanisms of clinically relevant microorganisms. *Clin Microbiol Rev.* 15: 167–193.

Donlan RM, Murga R, Carpenter J, Brown E, Besser R, Fields B. 2002. Monochloramine disinfection of biofilm-associated *Legionella pneumophila* in a potable water model system. *Legionella*. 406–410.

Donohue MJ, O'Connell K, Vesper SJ, Mistry JH, King D, Kostrich M, Pfaller S. 2014. Widespread molecular detection of *Legionella pneumophila* serotype 1 in cold water taps across the United States. *Environ Sci Technol.* 48: 3145–3152.

Falkinham JO III. 2015. Common features of opportunistic premise plumbing pathogens. *Int J Environ Res Public Health.* 12: 4533–4545.

Falkinham JO III, Pruden A, Edwards M. 2015. Opportunistic premise plumbing pathogens: Increasingly important pathogens in drinking water. *Pathogens.* 4: 373–386.

Falkinham JO III, Williams MD, Kwait R, Lande L. 2016. *Methylobacterium* spp. as an indicator for the presence or absence of *Mycobacterium* spp. *Intl J Mycobacteriol.* 5: 240–243.

Falkinham JO III. 2003. Factors influencing the chlorine susceptibility of *Mycobacterium avium, Mycobacterium intracellulare,* and *Mycobacterium scrofulaceum. Appl Environ Microbiol.* 69: 5685–5689.

Falkinham JO III. 2011. Nontuberculous mycobacteria from household plumbing of patients with nontuberculous mycobacteria disease. *Emerg Infect Dis.* 17: 419–424.

Falkinham JO III, Iseman MD, de Haas P, van Soolingen D. 2008. *Mycobacterium avium* in a shower linked to pulmonary disease. *J Water Health.* 6: 209–213.

Feazel LM, Baumgartner LK, Peterson KL, Frank DN, Harris JK, Pace NR. 2009. Opportunistic pathogens enriched in showerhead biofilms. *Proc Natl Acad Sci USA.* 106: 16393–16399.

Flint KP. 1987. The long term-survival of *Escherichia coli* in river water. *J Appl Bacteriol.* 63: 261–270.

Flourney DJ, Petrone RL, Voth DW. 1992. A pseudo-outbreak of *Methylobacterium mesophilica* isolated from patients undergoing bronchoscopy. *Euro J Clin Microbiol Infect Dis.* 11: 240–243.

Furuhata K, Banzai AU, Kawakami Y, Ishizaki N, Yoshida Y, Goto K, Fukuyama M. 2011. Genotyping and chlorine-resistance of *Methylobacterium aquaticum* isolated from water samples in Japan. *Biocontrol Science.* 16: 103–107.

Furuhata K, Koike KA, Matsumoto A. 1989. Growth and survival of a chlorine resistive Gram-negative rod bacterium, *Protomonas extorquens* isolated dominantly from drinking tank water. *Bull Jpn Soc Microbiol Ecol.* 4: 35–47.

Furuhata K, Koike KA. 1993. Isolation of *Methylobacterium* spp. from drinking tank water and resistance of isolates to chlorine. *Koshuu.* 40: 1047–1053. (In Japanese)

Gallego V, Garcia MT, Ventosa A. 2006. *Methylobacterium adhaesivum* sp. nov., a methylotrophic bacterium isolated from drinking water. *Int J Syst Evol Microbiol.* 56: 339–342.

Gilchrist MJ, Kraft JA, Hammond JG, Connelly BL, Myers MG. 1986. Detection of *Pseudomonas mesophilica* as a source of nosocomial infections in a bone marrow transplant unit. *J Clin Microbiol.* 23: 1052–1055.

Goeres DM, Loetterle LR, Hamilton MA, Murga R, Kirby DW, Donlan RM. 2005. Statistical assessment of a laboratory method for growing biofilms. *Microbiology.* 151: 757–762.

Gourion B, Francez-Charlot A, Vorholt J. 2007. PhyR is involved in the general stress response of *Methylobacterium extorquens* AM1. *J Bacteriol.* 190: 1027–1035.

Green PN, Ardley JK. 2018. Review of the genus *Methylobacterium* and closely related organisms: A proposal that some *Methylobacterium* species be reclassified into a new genus, *Methylorubrum* gen. nov. *Int J Syst Evol Microbiol.* 68: 2727–2748.

Green PN, Bousfield IJ. 1982. A taxonomic study of some Gram-negative facultatively methylotrophic bacteria. *Microbiol.* 128: 623–638.

Green PN. 2006. *Methylobacterium.* In: Dworkin M, Falkow S, Rosenberg E, Schleifer K-H, Stackebrandt E et al. (eds.). *The Prokaryotes.* Springer, New York, 257–265.

Hirashi A, Furuhata K, Matsumoto A, Koike KA, Fukuyama M, Tabuchi K. 1995. Phenotypic and genetic diversity of chlorine-resistant

Methylobacterium strains isolated from various environments. *Appl Environ Microbiol.* 2099–2107.

Holland MA, Polacco JC. 1994. PPFMS and other covert contaminants: Is there more to plant physiology than just plant? *Annu Rev Plant Physiol Plant Mol Biol.* 45: 197–209.

Hornei B, Lűneberg E, Schmidt-Rotte H, Maass M, Weber K, Heits F, Frosch M, Solbach W. 1999. Systemic infection of an immunocompromised patient with *Methylobacterium zatmanii. J Clin Microbiol.* 37: 248–250.

Hugenholtz P, Cunningham MA, Hendrlkz JK, Fuerst JA. 1995. Desiccation resistance of bacteria isolated from an air-handling system biofilm determined using a simple quantitative membrane filter method. *Lttrs Appl Microbiol.* 21: 41–46.

Imbert G, Seccia Y, La Scola B. 2005. *Methylobacterium* sp. bactaeremia due to contaminated endoscope. *J Hosp Infect.* 612: 268–270.

Ivanova EG, Doronina NV, Trotsenko YA. 2001. Aerobic methylobacteria are capable of synthesizing auxins. *Microbiologiya.* 70: 452–458.

Joe MM, Saravanan VS, Sa T. 2013. Aggregation of selected plant, growth-promoting *Methylobacterium* strains: Role of cell surface components and hydrophobicity. *Arch Microbiol.* 195: 219–225.

Kaneko M, Hiraishi A. 1991. Physico-chemical and microbiological research on pink-colored aggregates found in a potable water treatment and distribution system. *J Jpn Water Works Assoc.* 60: 23–34.

Kelley ST, Theisen U, Angenent LT, St Amand A, Pace NR. 2004. Molecular analysis of shower curtain biofilm microbes. *Appl Environ Microbiol.* 70: 4187–4192.

Kirschner RA Jr, Parker BC, Falkinham JO III. 1992. Epidemiology of infection of nontuberculous mycobacteria. X. *Mycobacterium avium, M. intracellulare,* and *M. scrofulaceum* in acid, brown swamps of the southeastern United States and their association with environmental variables. *Am Rev Respir Dis.* 145: 271–275.

Koeing RL, Morris RO, Polacco JC. 2002. tRNA is the source of low-level trans-zeatin production in *Methylobacterium* spp. *J Bacteriol.* 184: 1832–1842.

Kos J, Suskovic G, Vukovic S, Simpraga M, Frece J, Matosic S. 2003. Adhesion and aggregation ability of probiotic strain *Lactobacillus acidophilus* M92. *J Appl Micorbiol.* 94: 981–987.

Kovalea J, Degener JE, van der Mei H. 2014. *Methylobacterium* and its role in health care-associated infection. *J Clin Microbiol.* 52: 1317–1321.

Kovalea J. 2013. Microbiological safety in endoscope reprocessing. Ph.D. thesis. University of Groningen, Groningen, the Netherlands.

Kressel AB, Kidd F. 2001. Pseudo-outbreaks of *Mycobacterium chelonae* and *Methylobacterium mesophilicum* caused by contamination of an automated endoscopy washer. *Infect Control Hosp Epidemiol.* 22: 414–418.

La Duc MT, Dekas A, Osman S, Moissl C,Newcombe D, Venkateswaran K. 2007. Isolation and characterization of bacteria capable of tolerating the extreme conditions of clean room environments. *Appl Environ Microbiol.* 73: 2600–2611.

Lai CC, Cheng A, Liu WL, Tan CK, Huang YT, Chung KP, Lee MR, Hsueh PR. 2011. Infections caused by unusual *Methylobacterium* species. *J Clin Microbiol.* 49: 3329–3331.

Mora M, Perras A, Alekhova TA, Wink L, Krause R, Aleksandrova A, Novozhilova, Moissl-Eichinger C. 2016. Resilient microorganisms in dust samples of the International Space Station—survival of the adaptation specialists. *BioMed Central.* 4: 65.

Muńoz-Egea MC, Ji P, Pruden A, Falkinham JO III. 2017. Inhibition of adherence of *Mycobacterium avium* to plumbing surface biofilms of *Methylobacterium* spp. *Pathogens.* 6: 42.

Omer ZS, Tombolini R, Gerhardson B. 2004. Plant colonization by pink-pigmented facultative methylotrophic bacteria (PPFMs). *FEMS Microbiol Ecol.* 47: 319–326.

Orphan VJ, Taylor LT, Hafenbradl D, Delong EF. 1999. Culture-dependent and culture-independent characterization of microbial assemblages associated with high-temperature petroleum reservoirs. *Appl Environ Microbiol.* 66: 700–711.

Poirier A, Lapointe R, Claveau S, Joly JR. 1988. Bacteremia caused by *Pseudomonas mesophilica. Can Med Assoc J.* 139: 411–412.

Raja P, Balachandar D, Sundaram SP. 2008. Genetic diversity and phylogeny of pink-pigmented facultative methylotrophic bacteria isolated from the phyllosphere of tropical crop plants. *Biol Fertil Soils.* 45: 45–53.

Rutherford PC, Narkowicz JE, Wood CJ, Peel MM. 1988. Peritonitis caused by *Pseudomonas mesophilica* in a patient undergoing continuous ambulatory peritoneal dialysis. *J Clin Microbiol.* 26: 2441–2443.

Sanders JW, Martin JW, Hooke M, Hooke J. 2000. *Methylobacterium mesophilicum* infection: Case report and literature review of an unusual opportunistic pathogen. *Clin Infect Dis.* 30: 936–938.

Schauer S, Kutschera U. 2011. A novel growth-promoting microbe, *Methylobacterium funariae* sp. nov., isolated from the leaf surface of a common moss. *Plant Signal Behav.* 6: 510–515.

Schulze-Röbbecke R, Buchholtz K. 1992. Heat susceptibility of aquatic mycobacteria. *Appl Environ Microbiol.* 58: 1869–1873.

Simmon RB, Rose LJ, Crow SA, Ahearn DG. 1999. The occurrence and persistence of mixed biofilms in automobile air conditioning systems. *Current Microbiol.* 39: 141–145.

Simoes LC, Simoes M, Vieira MJ. 2009. Susceptibility of drinking water biofilm bacteria to sodium hypochlorite. 124–125. In Eurobiofilms Book of abstracts. http://repositorium.sdum.uminho.pt/bitstream/1822/9607/1/EUROBIOFILMS2009_Abstract-Simoes%5B1%5D.pdf.

Smith SM, Eng RHK, Forrester C. 1985. *Pseudomonas mesophilica* infections in humans. *J Clin Microbiol.* 21: 314–317.

Soucie WJ, Schuler B. 2006. Avoiding pink stain pain. *American Water Works Assoc.* 26–29.

Sy A, Giraud E, Jourand P. 2001. Methylotrophic *Methylobacterium* bacteria nodulate and fix nitrogen in symbiosis with legumes. *J Bacteriol.* 183: 214–220.

Thomas V, Herrera-Rimann K, Blanc DS, Greub G. 2006. Biodiversity of amoebae and amoebae-resisting bacteria in a hospital water network. *Appl Environ Microbiol.* 72: 2428–2438.

Tsagkari E, Keating C, Couto JM, Sloan WT. 2017. A Keystone *Methylobacterium* strain in biofilm formation in drinking water. *Water.* 9: 778. doi: 10.3390/w9100778.

Van Loosdrecht MC, Lyklema J, Norde W, Schraa G, Zehnder AJ. 1987. The role of bacterial cell wall hydrophobicity in adhesion. *Appl Environ Microbiology.* 53: 1893–1897.

Verginer M, Siegmund B, Cardinale M, Müller H, Choi Y, Míguez CB, et al. 2010. Monitoring the plant epiphyte *Methylobacterium extorquens* DSM 21961 by real-time PCR and its influence on the strawberry flavor. *FEMS Microbiol Ecol.* 74: 136–145. doi: 10.1111/j.1574-6941.2010.00942.x

Wasserman B, Műller H, Berg G. 2019. An Apple a Day: Which bacteria Do we eat with organic and conventional apples? *Front Microbiol.* 10: 1629. doi.org/10.3389/fmicb.2019.01629

Waturangi DA, Kusuma A. 2008. Analysis of pink-pigmented facultative methylotroph bacteria from human environments. *Microbiol Indones.* 2: 112–114.

Webster E, Ribner B, Streed LL, Hutton N. 1996. Microbial contamination of activated 2% glutaraldehyde used in high-level disinfection of endoscopes. *Am J Infect Control.* 24: 153.

Yano T, Kubota H, Hanai J, Hitomi J, Tokuda H. 2013. Stress tolerance of *Methylobacterium* biofilms in bathrooms. *Microbes Environ.* 28: 87–95.

7.5 *Segniliparus* Species

7.5.1 Introduction

Segniliparus spp. were introduced to me through a parent of twin boys with cystic fibrosis and *Segniliparus* spp. pulmonary disease. My lab tried to recover *Segniliparus* spp. isolates from water and soil samples collected from the home of the twins; we did not isolate any *Segniliparus* spp. strains. As is the case for a number of the newly emerging opportunistic premise plumbing pathogens (OPPPs), little in the way of systematic thorough environmental studies have been performed. The characteristics and taxonomy of *Segniliparus* spp. have been described (Butler et al., 2005) and the genome sequences of *Segniliparus rotundus* (Sikorski et al., 2010) and *Segniliparus rugosus* (Earl et al., 2011) have been published.

7.5.2 Disease and Risk Factors

Clearly, cystic fibrosis is a risk factor for infection with *Segniliparus* spp., as the first isolations of *S. rugosus* were from twin boys with cystic fibrosis (Butler et al., 2007). Further, an Australian teenaged female from the tropical north of Queensland with cystic fibrosis was diagnosed with *S. rugosus* infection (Hansen et al., 2009).

S. rotundus was isolated from sputum samples from a 43-year old woman with bronchiectasis and history of tuberculous pleurisy and a 3-month history of a chronic cough and sputum production (Koh et al., 2011). It is not clear whether bronchiectasis was a predisposing factor for *S. rotundus* infection or a consequence.

7.5.3 Unique *Segniliparus* spp. Features

A unique feature of members of *Segniliparus* spp. is their novel and very long-chain (C_{100}) mycolic acids (Butler et al., 2005; Hong et al., 2012; Lanéelle et al., 2013) that bear fundamental differences with the mycolic acids of members of the genus *Mycobacterium* (Hong et al., 2010). Like the *Mycobacterium* spp., mycolic acids fold and form a hydrophobic, impermeable outer membrane (Hong et al., 2012).

7.5.4 Ecology and Sources of *Segniliparus* spp.

Although not a human infection case, *S. rugosus*-associated bronchiolitis was described in a California Sea Lion (Evans, 2011). That report suggests that *Segniliparus* spp. maybe adapted to an estuarine habitat, as are members of the *Mycobacterium avium* complex (George et al., 1980).

References

Butler WR, Floyd MM, Brown JM, Toney SR, Daneshvar MI, Cooksey RC, Carr J, Steigerwalt AG, Charles N. 2005. Novel mycolic acid-containing bacteria in the family *Segniliparaceae* fam. Nov., including the genus *Segniliparus* gen. nov., with descriptions of *Segniliparus rotundus* sp. nov. and *Segniliparus rugosus* sp. nov. *Int J Syst Evol Microbiol.* 55: 1615–1624.

Butler WR, Sheils CA, Brown-Elliott BA, Charles N, Colin AA, Gant MJ, Goodill J, Hindman D, Toney SR, Wallace RH Jr, Yakrus MA. 2007. First isolations of *Segniliparus rugosus* from patients with cystic fibrosis. *J Clin Microbiol.* 45: 3449–3452.

Earl AM, Desjardins CA, Fitzgerald MG, Arachchi HM, Zeng Q, Mehta T, Griggs A, Birren BW, Toney NC, Carr J, Posey J, Butler WR. High quality draft genome sequence of *Segniliparus rugosus* CDC 945^T=(ATCC BAA-974^T). *Std Genom Sci.* 5: 389–397.

Evans RH. 2011. *Segniliparus rugosus*-associated bronchiolitis in a California sea lion. *Emerg Infect Dis.* 17: 311–312.

George KL, Parker BC, Gruft H, Falkinham JO III. 1980. Epidemiology of infection by nontuberculous mycobacteria. II. Growth and survival in natural waters. *Am Rev Respir Dis.* 122: 89–94.

Hansen T, Van Kerckhof J, Jelfs P, Wainwright C, Ryan P, Coulter C. 2009. *Segniliparus rugosus* infection in Australia. *Emerg Infect Dis.* 15: 611–613.

Hong S, Cheng T-Y, Layre E, Sweet L, Young D, Posey JE, Butler WR, Moody DB. 2012. Ultralong C_{100} mycolic acids support the assignment of *Segniliparus* as a new bacterial genus. *PLoS ONE.* 7: E39017.

Koh W-J, Choi G-E, Lee S-H, Park YK, Lee NY, Shin SJ. 2011. First case of *Segniliparus rotundus* pneumonia in a patient with bronchiectasis. *J. Clin Microbiol.* 49: 3403–3405.

Lanéelle M-A, Eynard N, Spina L, Lemassu A, Laval F, Huc E, Etienne G, Marrakchi H, Daffé. 2013. Structural elucidation and genomic scrutiny of the C_{60}–C_{100} mycolic acids of *Segniliparus rotundus*. *Microbiol.* 159: 191–203.

Sikorski J, Lapidus A, Copeland A, Misra M, et al. 2010. Complete genome sequence of *Segniliparus rotundus* type strain CDC (1076^T). *Std Genom Sci.* 2: 203–211.

7.6 *Cupriavidus* spp.

7.6.1 Introduction

Cupriavidus spp. are newly emerging opportunistic premise plumbing pathogens (OPPPs). The current lack of a review article describing their epidemiology is due, in part, to the changing genus name for this group of bacteria, namely R*alstonia, Wauteria,* and *Cupriavidus.* My first introduction to these bacteria was through Dr. Earl Casida, a soil microbiologist at Penn State. He was isolating, studying, and characterizing predator bacteria, which were capable of killing and parasitizing other, living bacteria, and fungi. One predator, *Cupriavidus mecator* was studied in detail (Makkar and Casida, 1987). In the absence of systematic studies of their geographic distribution, all that can be written is that they have been isolated from both natural and engineered water systems. For the purpose of this volume's focus on OPPPs, *Cupriavidus* spp. are present in premise plumbing. As listed below, there are a variety of presentations for *Cupriavidus* spp. infections, but there are a few investigations of *Cupriavidus* taxonomy and the identification of sources of the infection.

7.6.2 Disease and Risk Factors

Reduced immunocompetence appears to be a risk factor for *Cupriavidus* spp. infection. The earliest report of a *Cupriavidus* spp. (nee *Ralstonia gilardii*) infection was in 2001 that described catheter-associated sepsis in a child with acute lymphoblastic leukemia (Wauters et al., 2001). *Cupriavidus gilardii* was reported as a cause of fatal infection in a 12-year old with aplastic anemia (Karafin et al., 2010). Immunodeficiency was ascribed as responsible for a *C. gilardii* muscle abscess infection in a renal transplant recipient (Tena et al., 2014). Cystic fibrosis is another risk factor for *Cupriavidus* spp. infection. *Cupriavidus* spp., *Ralstonia pickettii,* or *Ralstonia mannitolilytica* pulmonary infection was reported in patients with cystic fibrosis (Coenye et al., 2002; Kalka-Moll et al., 2009). *Cupriavidus* spp. infections have been reported in elderly, non-immunocompromised patients as well. One case report described a 98-year-old woman with a *C. gilardii* pacemaker-associated bacteremia (Kobayashi et al., 2016) and the second reported a C. gilardii bacteremia in an 87-year old man with chronic obstructive pulmonary disease (Zhang et al., 2017). A third case involved a 74-year old man with type-2 diabetes, arteriosclerotic heart disease, dyslipidemia, arterial hypertension, and obesity; after surgery for prostate cancer, who was shown to be infected with *Cupriavidus metallidurans*

(Langevin et al., 2011). Most recently, there is a report of an outbreak of *Cupriavidus pauculus* bloodstream infections in patients in a bone marrow transplant unit (Health Protection Scotland, 2018). Water samples were collected within the unit and *C. pauculus* was isolated from one sink in the unit. In addition to the removal of the sink, patients were moved from the bone marrow transplant unit to other wards in the hospital (Health Protection Scotland, 2018).

7.6.3 Epidemiology and *Cupriavidus* Infection Sources

Based on the report from Health Protection Scotland, it can be assumed that one source of *Cupriavidus* sp. infections in hospitals is water, taps, sinks, and drains (Health Protection Scotland, 2018). However, as the taxonomy of *Cupriavidus* spp. has changed (Vandamme and Coeyne, 2004), it has been difficult to find information on other sources of this genus. In addition to the Health Protection Scotland report, other publications document hospital water as a *Cupriavidus* spp. source of infection. Pseudo-infections by *C. pauculus* in an outpatient clinic were traced by matching PFGE-patterns to tap water used to rinsing culture swabs in tap water (Balada-Llasat et al., 2010). A cluster of four infants infected with *Cupriavidus pauculus* undergoing extracorporeal membrane oxygenation (ECMO) was traced to the water reservoirs of the ECMO instrument; however, no fingerprint comparisons were performed to prove that the *C. pauculus* isolates from the water reservoirs shared the same fingerprint pattern as those of the infected patients (Stovall et al., 2010).

7.6.4 Unique *Cupriavidus* Structural Features

Members of *Cupriavidus* spp. are possibly all metal-resistant, although not every species has been specifically tested. The genus name is based on its resistance to copper and the Cu-responsive growth (Makkar and Casida, 1987; Nies et al., 2006). In addition to copper resistance, *Cupriavidus necator* was shown to kill surrounding bacterial cells by actively extracting copper; their predator phenotype (Makkar and Casida, 1987). Plasmids encoding metal resistance have been isolated and described (Monchy et al., 2007; Nies et al., 2006). It is quite possible that a copper-based selective agar medium could be developed for *Cupriavidus* spp.

Other characteristics of *Cupriavidus* spp. include the fact they are chemolithotrophic, thereby demonstrating that they are not obligate predators of other microorganisms (Sato et al., 2006). *Cupriavidus* spp. strains are capable of hydrogen-oxidation (Sato et al., 2006), although the significance of that observation awaits further investigation. *Cupriavidus* spp. have also been isolated as root nodule plant symbionts (Barrett and Parker, 2006), suggesting an adaptation to habitats other than water.

7.6.5 Ecology of *Cupriavidus* spp.

It is likely, although not systematically studied that *Cupriavidus* spp. are widely distributed in the environment, where they have opportunities to come in contact with humans and plants. *Cupriavidus* spp. have been recovered from both water and soil (Coenyne et al., 2002; Sato et al., 2006). Thus, they should not be considered metal-resistant bacteria and limited to a narrow range of metal-contaminated habitats. Although their growth is stimulated by copper, they can be grown in the absence of copper (Falkinham, unpublished). The copper resistance of *Cupriavidus metallidurans* was shown to mediate the long-term survival of the bacterium in copper pipes (Maertens et al., 2020), thus documenting its role as a determinant of its ecology and habitats. That knowledge raises the question of the use of copper pipe to reduce numbers of OPPPs in premise plumbing; it may select for *Cupriavidus* spp.

References

Balada-Llasat J-M, Elkins C, Swyers L, Bannerman T, Pancholi P. 2010. Pseudo-Outbreak of *Cupriavidus pauculus* infection at an outpatient clinic related to rinsing culturette swabs in tap water. *J Clin Microbiol.* 48: 2645–2647.

Barratt CF, Parker MA. 2006. Coexistence of *Burkholderia, Cupriavidus,* and *Rhizobium* sp. nodule bacteria on two *Mimosa* spp. in Costa Rica. *Appl Environ Microbiol.* 72: 1198–1206.

Coenye T, Vandamme P, LiPuma JJ. 2002. Infection by *Ralstonia* species in cystic fibrosis patients: Identification of R. *pickettii* and *R. mannitolytica* by polymerase chain reaction. *Emerg Infect Dis.* 8: 692–696.

Health Protection Scotland. 2018. Summary of incident and findings of the NHS Greater Glasgow and Clyde: Queen Elizabeth University Hospital/Royal Hospital for Children water contamination incident and recommendations for NHS Scotland.

Kalka-Mill WM, LiPuma JJ, Acurso FJ, Plum G, van Konningsbruggen S, Vandamme P. 2009. Airway infection with a novel *Cupriavidus* species in persons with cystic fibrosis. *J. Clin Microbiol.* 47: 3026–3028.

Karafin M, Romagnoli M, Fink DL, Howard T, Rau R, Milstone AM, Carroll KC. 2010. Fatal infection caused by *Cupriavidus gilardii* in a child with aplastic anemia. *J. Clin Microbiol.* 48: 1005–1007.

Kobayashi T, Nakamura I, Fujita H, Tsukimori A, Sato A, Fukushima S. 2016. First case report of infection due to *Cupriavidus gilardii* in a patient without immunodeficiency: A case report. *BMC Infect Dis.* 16: 493.

Langevin S, Vincelette J, Bekal S, Gaudreau C. 2011. First case of invasive human infection caused by *Cupriavidus metallidurans. J Clin Microbiol.* 49: 744–745.

Maertens L, Coninx I, Claesen J, Leys N, Matroule J-Y, Van Houdt R. 2020. Copper resistance mediates long-term survival of *Cupriavidus metallidurans* in wet contact with metallic copper. *Front Microbiol.* 11: 1208. Doi: 10.3389/fnucb,2020.01208.

Makkar NS, Casida LE Jr. 1987. *Cupriavidus necator* gen. nov., sp. nov.: A nonobligate bacterial predator of bacteria in soil. *Int J System Bacteriol.* 37: 323–326.

Monchy S, Benotmane MA, Janssen P, Vallaeys T, Taghavi S, van der Lelie D, Mergeay M. 2007. Plasmids pMOL28 and pMOL30 of *Cupriavidus metallidurans* are specialized in the maximal viable response to heavy metals. *J Bacteriol.* 189: 7417–7425.

Nies DH, Rehbein G, Hoffmann T, Baumann C, Grosse C. 2006. Paralogs of genes encoding metal resistance proteins in *Cupriavidus metallidurans* strain CH34. *J Mol Microbiol Biotechnol.* 11: 82–93.

Sato Y, Nishihara H, Yoshida M, Watanabe M, Rondal JD, Conception RN, Ohta H. 2006. *Cupriavidus pinatubonensis* sp. nov. and *Cupriavidus laharis* sp. nov., novel hydrogen-oxidizing, facultatively chemolithotrophic bacteria isolated from volcanic mudflow deposits from Mt. Pinatubo in the Philippines. *Int J Syst Evol Microbiol.* 56: 973–978.

Stovall SH, Wisdom C, McKamie W, Ware W, Dedman H, Fiser RT. 2010. Nosocomial transmission of *Cupriavidus pauculus* during extracorporeal membrane oxygenation. *ASAIO J.* 56: 486–487.

Tena D, Losa C, Medina MJ, Saez-Nieto JA. 2014. Muscular abscess caused by *Cupriavidus gilardii* in a renal transplant recipient. *Diagn Microbiol Infect Dis.* 79: 108–110.

Vandamme P, Coenye T. 2004. Taxonomy of the genus *Cupriavidus*: A tale of lost and found. *Int J System Evol Microbiol.* 54: 2285–2289.

Wauters G, Claeys G, Verschraegen G, DeBaere T, Vandecruys E, Van Simaey L, DeGanck C, Veneechoutte M. 2001. Case of catheter sepsis with *Ralstonia gilardii* in a child with acute lymphoblastic leukemia. *J Clin Microbiol.* 39: 4583–4584.

Zhang Z, Deng W, Wang S, Xu L, Yan L, Liao P. 2017. First case report of infection caused by *Cupriavidus gilardii* in a non-immunocompromised Chinese patient. *ID Cases.* 10: 127–129.

Chapter 8

Factors Selecting for Opportunistic Premise Plumbing Pathogens in Premise Plumbing

8.1 Introduction

The opportunistic pathogens are normal inhabitants of premise plumbing; the water pipes in houses, apartments, hospitals, healthcare facilities, and any building. These opportunistic premise plumbing pathogens (OPPPs) include *Legionella pneumophila, Mycobacterium avium* complex (MAC), *Pseudomonas aeruginosa, Stentrophomonas maltophila, Acinetobacter baumannii, Sphingomonas paucimobilis, Methylobacterium* spp., *Segniliparus* spp., *Cupriavidus* spp., and amoebae. All are highly adapted to persist and grow in premise plumbing. The physiochemical factors leading to their residence in premise plumbing include disinfectant resistance, surface adherence and biofilm formation, growth at low organic carbon levels (oligotrophy), growth at low oxygen concentrations (stagnation), heat resistance, and desiccation tolerance (Table 8.1). In addition, genetic factors leading to genome reorganization appear to drive OPPP adaptation to survival, growth, and persistence in the human environment as well as drive increased virulence (Rodriguez-Rojas et al., 2012). As the incidence of OPPP disease is increasing and standard

Opportunistic Premise Plumbing Pathogens
Joseph O. Falkinham, III

ISBN 978-981-4968-40-9 (Hardcover), 978-1-003-32100-2 (eBook)
www.jennystanford.com

water treatment technologies actually select for OPPPs, the water industry is challenged to develop new control strategies.

Table 8.1 Physiochemical factors influencing OPPP numbers in premise plumbing

Disinfectant Resistance
Surface Adherence
Biofilm Formation
Growth at Low Organic Carbon levels (Oligotrophy)
Growth at Low Oxygen Concentrations
Water Age
Heat Resistance
Desiccation Tolerance
Growth and Survival in Amoebae
Adaptation to Stress

8.2 Physiochemical Factors Selecting for OPPPs in Premise Plumbing

8.2.1 Surface Adherence and Biofilm Formation

Premise plumbing has a number of unique characteristics to which the OPPPs are highly adapted. First, premise plumbing is a flowing system and in the absence of surface attachment, any microorganism will be washed out. Thus, adherence and biofilm formation will lead to the persistence and even growth if nutrient and conditions are suitable. Fortunately, premise plumbing offers a high surface-to-volume ratio (Chapter 3), providing a very large surface area for adherence and biofilm formation. All the OPPPs form biofilms; for example, *L. pneumophila* (Bezanson et al., 1992; Rogers et al., 1994); *P. aeruginosa* (Spoering and Lewis, 2001), MAC (Mullis and Falkinham, 2013), *S. maltophilia* (Ryan et al., 2009), *A. baumannii* (Ciofu and Tolker-Nielsen, 2019), and *Methylobacterium* spp. (Yabu et al., 2013). Due to the reduced penetration of chlorine (De Beer et al., 1994a) and oxygen (DeBeer et al., 1994b), biofilm formation has a major effect on disinfection and growth. Biofilm formation is especially important for the survival of OPPPs in a flowing system, as they are relatively slow-growing compared to fecal contaminants of water like *E. coli*. In fact, the high surface hydrophobicity of the OPPPs promotes adherence (Absolom, 1988). Thus, premise plumbing selects for OPPPs through biofilm formation.

8.2.2 Disinfectant Resistance

Drinking water is treated water with low concentrations of residual disinfectant (e.g., chlorine, chloramine, or ozone). Thus, there is a selection for disinfectant-resistant. The OPPPs are disinfectant-resistant, including *L. pneumophila* (Kuchta et al., 1985), *P. aeruginosa* (Grobe et al., 2001), MAC (Taylor et al., 2000), *A. baumannii* (Karamathil et al., 2014), and *Methylobacterium* spp. (Hiraishi et al., 1995). By contrast, in premise plumbing with residual disinfectant, the fecal pathogens, such as *E. coli*, *Salmonella*, and *Shigella*, will be killed, leaving only the OPPPs. Thus, disinfection provides a selective environment, leading to the death of fecal coliforms and their replacement by disinfectant-resistant OPPPs. For example, *M. avium* is 2,000-fold more tolerant of 1 ppm chlorine compared to *E. coli*; the $CT_{99.9\%}$ for *M. avium* is approximately 100, while the $CT_{99.9\%}$ for *E. coli* is 0.05 (Taylor et al., 2000).

Biofilm formation adds a second layer of defense against disinfectants, as the layers of cells and extracellular matrix in biofilms increase resistance to disinfectants by serving as a permeability barrier (De Beer et al., 1994a). *M. avium* cells in biofilms are more resistant to chlorine than suspended cells (Steed and Falkinham, 2006) as are cells of *L. pneumophila* (Rogers et al., 1994) and *P. aeruginosa* (Spoering and Lewis, 2001). Further, *M. avium* cells grown in biofilms, yet isolated and placed in single-cell suspension, are more resistant to disinfection (Steed and Falkinham, 2006). This increased disinfectant resistance is evidently an adaptation as those resistant, biofilm-grown cells regain the disinfectant-susceptibility of medium-grown, single cells (Steed and Falkinham, 2006).

8.2.3 Low Organic Matter Concentration

Treated water in premise plumbing has a relatively low concentration of organic matter, expressed as assimilable organic carbon (AOC). That is thought to limit the growth of pathogens and other microorganisms in drinking water. Although microbiologists grow OPPPs in a rich laboratory medium for detection, enumeration, and identification, OPPPs can grow at low concentrations of organic carbon. OPPPs behave like oligotrophs, able to grow on low organic carbon. For example, *P. aeruginosa* has been shown able to grow in distilled water (Favero et al., 1971) and a strain of *M. avium* persisted in a pilot drinking water system at AOC levels as low as 50 μg/L (Norton et al., 2004).

8.2.4 Stagnation and Low Oxygen

Low oxygen levels occur in premise plumbing during periods of no flow (e.g., stagnation), yet they do not appear to limit the growth of OPPPs. In fact, *P. aeruginosa* is capable of growth under anaerobic conditions by utilizing nitrate as a terminal electron acceptor in the absence of oxygen (Palmer et al., 2007). Further, although MAC bacteria are lung pathogens and thereby considered obligate aerobes that are not the case. MAC strains have been shown to grow rapidly at 12% oxygen, compared to their growth in the air (21% oxygen) (Lewis and Falkinham, 2015). Further, growth even occurs at 6% oxygen, albeit at half the rate as cells grow in the air or 12% oxygen [20]. Thus, OPPPs should be considered micro aerobes.

The preference for biofilm residence by the OPPPs also contributes to the selection for the growth at low oxygen. The penetration of oxygen to cells in biofilms is reduced (De Beer et al., 1994b).

Water age, a term representing the length of time water is resident in a water system, is another factor influencing OPPP numbers in premise plumbing. Water age in a house is typically 3–4 days. However, in the absence of water movement, due to no usage or abandonment of a portion of the premise plumbing, the water age can be increased. As water age increases, microbial numbers increase. Although the oxygen content of such stagnant water will slowly fall, we have seen that reduced oxygen levels will not particularly limit OPPPs' growth. They will grow, albeit slowly. Stagnation has become a major topic of conversation in the water and building-management industries, due to the COVID-19 pandemic. Office buildings have been closed, leading to dramatic increases of *L. pneumophilia* and other OPPPs in building water. One response has been to urge building managers to flush water through the buildings to reduce water age.

The extreme version of the increased water age is a dead-end in plumbing. There is no water circulation in a dead-end; water enters but does not come out. Consequently, high OPPP numbers can be found in dead-ends. Even when measures are taken to disinfect premise plumbing, such as hyperchlorination or elevated temperature, none of the treated water enters the dead-end. Consequently, following any type of disinfection regimen, OPPPs in a dead-end can re-inoculate the system, resulting in the re-appearance of OPPPs throughout premise plumbing.

8.2.5 Amoebae Predation

It is well-established that amoebae graze on microorganisms in premise plumbing as a source of nutrients for growth (Thomas et al., 2006; Lau and Ashbolt, 2009). However, rather than be phagocytized and killed by amoebae as other waterborne bacteria, OPPPs survive and grow in amoebae. *L. pneumophila* (Lau and Ashbolt, 2009), *M. avium* (Cirillo et al., 1997), *P. aeruginosa* (Dey et al., 2019), *Methylobacterium* spp. (Thomas et al., 2006), *A. baumannii* (Cateau et al., 2011), and *S. maltophilia* (Cateau et al., 2014) have all been shown to survive and grow in amoebae. A current name for such bacteria is amoebae-resisting microorganisms (ARMs), and it is apparent that amoebae serve as selective agents for the OPPPs in drinking water distribution systems and premise plumbing.

Beyond supporting the growth and persistence of OPPPs in drinking water, it is quite possible that amoebae may maintain the viability and growth of some OPPPs, like *L. pneumophila*. Amoebae can resuscitate viable, but unculturable (VBNC) *L. pneumophila* (Li et al., 2014). Further, it is possible to enrich water samples for OPPPs by adding amoebae, incubating, and harvesting amoebae (Falkinham, unpublished).

8.2.6 Heat Tolerance

Although not all OPPPs have been studied in-depth, many appear to be resistant to high temperatures and able to persist in hot water systems in hospitals and homes, e.g., *L. pneumophila* (Schofield and Locci, 1985; Arnow et al., 1985) and *M. avium* (Falkinham, 2011; Guenette et al., 2020).

Recently, my colleagues and I have become interested in the heat tolerance of *M. avium*. That interest was triggered by several observations. First, *Mycobacterium* spp. numbers were higher in homes of patients whose water heater was set to 125 °F (50 °C) or below compared to patients whose water heater was set to 130 °F (55 °C) whose plumbing rarely had *Mycobacterium* spp. (Falkinham, 2011). Second, based on studies of household plumbing of *M. avium*-infected patients in Wynnewood, PA, a suburb of Philadelphia, it appeared that the household water heaters were sites for the amplification of *M. avium* numbers. Water from the water heater had significantly more *M. avium* cells than cold water that is the source of water in the water heater (Lande et al., 2019). On the basis of those discoveries, the water heater temperature was raised in a proportion of homes, and within 8–12 weeks, *M. avium* had disappeared. As part of that remediation project, isolates of *M. avium* were recovered from the patient's premise plumbing to make sure that the elevation in water heater temperature was not selected for temperature-resistant mutants. We were surprised that the isolates were more tolerant of high temperatures than reported in the literature (Schulze-Röbbecke and Buchholtz, 1992). Specifically, a 3-hour exposure of *M. avium* cells to 50 °C, 55 °C, or 60 °C, failed to reduce colony counts. Complete survival of *M. avium* cells exposed to 60 °C for 3 h would not be expected based on published data (Schulze-Röbecke and Buchholtz, 1992). Only exposure of cells of different patients and household *M. avium* isolates to 65 °C led to a significant loss of colony count (Falkinham, 2020). Examination of the *M. avium* isolates recovered following the increase in water heater temperature, and revealed that the growth temperature directly influenced survival to 65 °C. The isolates from the houses after the increased water heater temperature was not due to the mutation, but to a pre-existing ability of adaptation. Cells grown at 42 °C, could survive 65 °C, but cells grown at 37 °C did not (Guenette et al., 2020). It demonstrated that *M. avium* and by extension to perhaps other OPPPs are capable of adaptation to high-temperature tolerance.

8.2.7 Desiccation Resistance of OPPPs

A major reason for including some of the emerging OPPPs in this volume is that those species have novel characteristics that have not been investigated in all OPPPs, but are logical traits providing a selective advantage for premise plumbing pathogens. One of those traits I discovered in reviewing the literature on *Methylobacterium*. *Methylobacterium* is relatively desiccation-resistant (Yano et al., 2013). The survival benefit of desiccation resistance is under circumstances where water flow is intermittent and the biofilms are subject to drying.

Recently, we have discovered that *Mycobacterium chimaera*, a member of the MAC is desiccation-resistant, as other members of MAC. The question of the desiccation tolerance of *M. chimaera* rose as a result of studies of the behavior of *M. chimaera* in heater-coolers. Heater-coolers are hospital, operating room devices that cool patients and keep blood warm while outside the body for oxygen-carbon dioxide exchange during cardiac surgery when the heart is stopped. *M. chimaera* infections were discovered in patients following cardiac surgery, and the bacterium traced to the water reservoirs of Sorin 3T heater-coolers (Sax et al., 2015). The water reservoirs of Sorin 3T heater-coolers were found to be colonized by the same strain of *M. chimaera* that was introduced into the device during functional testing of the instrument after assembly. As tap water used in a factory in Munich, Germany, carried *M. chimaera*, that OPPP was present in many Sorin 3T instruments, and infections were reported throughout the world. As the instruments were filled with Munich water, tested, and drained before shipping, the question was how did the *M. chimaera* strain survive the ensuing desiccation during shipping? For example, shipping to the United States took approximately 3 weeks. Measurements of survival of water-acclimated cells of two independent *M. chimaera*-Sorin 3T isolates showed that *M. chimaera* is desiccation-tolerant with

40–50% survival after 42-days of incubation of cells on stainless steel surfaces (Falkinham, in preparation).

Those results and the reports on desiccation tolerance of *Methylobacterium* spp. have led to the addition of desiccation tolerance to my list of common characteristics of OPPPs.

8.3 Genetic Factors Influencing OPPPs in Premise Plumbing

8.3.1 Genome Reorganization

It was proposed that genome reorganization is responsible for the persistence and virulence of *Pseudomonas aeruginosa* in cystic fibrosis patients (Rodriguez-Rojas et al., 2012). The hypothesis is that genome rearrangements are constantly occurring and subject to strong selection in the human lung during infection. A corollary of the hypothesis is that infecting isolates with high rates of genome rearrangement will eventually predominate in the lung in response to selection. Isolates with low rates of genome rearrangement will not persist as they do not generate sufficient numbers of mutant variants to ensure survival, growth, and persistence. This is due to the fact that the source of *P. aeruginosa, A. baumannii,* and *M. avium* is drinking water in the human environment. Infecting isolates of those OPPPs predominate in drinking water as a result of selection in that low nutrient habitat.

Today, there is strong evidence for genome evolution from analysis of whole-genome sequences of serial isolates of *P. aeruginosa* (Bianconi et al., 2019; Kidd et al., 2012) and *Acinetobacter baumannii* (Bogaty et al., 2018) isolates from infected patients. Genome reorganization has become apparent during the course of infection in humans; particularly in *Pseudomonas aeruginosa* and *Burkholderia cepacia* (Govan et al., 2007; Wargo, 2019). That was discovered by following patients (e.g., cystic fibrosis) for a long time by repeated culture over time and trying to discover whether recurrent infections were due to "reactivation" of the original infecting strain or due to infection by another strain. That work was done by DNA fingerprinting and now by whole-genome analysis. It would be understood that sequential genome analysis of an OPPP could lead to the identification of virulence genes.

8.3.2 Horizontal Gene Transmission

Both *P. aeruginosa* and *B. cepacia* have active horizontal gene transfer mechanisms. Those mechanisms involve the transfer of plasmids and transposable genetic elements that are drivers of chromosome rearrangements. *Acinetobacter baumannii, Stenotrophomonas maltophilia,* and *Sphingomonas paucimobilis,* which were originally considered pseudomonads, share genetic determinants of horizontal gene transfer. Horizontal genetic transmission, along with transposition of insertion sequences (IS) and changes in copy numbers in tandem duplications used in fingerprinting isolates are mechanisms of chromosome reorganization.

8.3.3 Survival Benefit of Genome Reorganization

Now, the question is why? What is the survival benefit? I think that chromosome reorganization is of adaptive survival benefit for bacteria that are free-living in water and soil. Bacteria have small genomes with very little extra DNA. Dividing the genome size into base pairs by 1,000 gives a good estimate of the number of genes. That is not the case in other organisms. Thus, without horizontal gene transfer, changing environmental conditions would mean the death of many bacteria.

I would suggest that the bacterial microbiome of aquatic bacteria, namely the Gram-negative with their outer membrane, serves as an Amazon of genes. A high level of horizontal gene transmission allows microbes to survive a changing environment by the acquisition of genes from other bacteria.

Many other bacterial pathogens do not face a changing fluctuating environment; they are commensals of humans (e.g., *Staphylococcus aureus* or *Streptococcus pneumoniae*) or obligate pathogens (e.g., *Mycobacterium tuberculosis*) and rely on humans to provide them with a stable environment.

References

Absolom DR. 1988. The role of bacterial hydrophobicity in infection: Bacterial adhesion and phagocytic ingestion. *Can J Microbiol.* 34: 287–298.

Arnow PM, Weil D, Para MF. 1985. Prevalence and significance of *Legionella pneumophila* contamination of residential hot-tap water systems. *J Infect Dis.* 152: 145–151.

Bezanson G, Burbridge S, Haldane D, Marrie T. 1992. In situ colonization of polyvinyl chloride, brass, and copper by *Legionella pneumophila. Can J Microbiol.* 38: 328–330.

Bianconi I, D'Arcangelo S, Esposito A, Benedet M, Piffer E, Dkinnella G, Gualdi P, Schinella M, Baldo E, Donati C, Jousson O. 2019. Persistence and microevolution of *Pseudomonas aeruginosa* in the cystic fibrosis lung: A single patient longitudinal genomic study. *Front Microbiol.* 9: 3242. doi: 10.3389/fmicb.2018.0342.

Bogaty C, Mataseje L, Gray A, Lefebvre B, Lévesque S, Mulvey M, Longtin Y. 2018. Investigation of a carbapenemase-producing *Acinetobacter baumannii* outbreak using whole genome sequencing versus a standard epidemiologic investigation. *Antimicrob Resist Infect Control.* 7: 140.

Cateau E, Maisonneuve E, Peguilhan S, Quellard N, Hechard Y, Rodier M-H. 2014. *Stentrophomonas maltophila* and *Vermamoeba vermiformis* relationships: Bacterial multiplication and protection in amoebal-derived structures. *Res Microbiol.* 20: 1–5.

Cateau E, Verdon J, Fernandez B, Hechard Y, Rodier M-H. 2011. *Acanthamoeba* sp. promotes the survival and growth of *Acinetobacter baumanii. FEMS Microbiol Lett.* 319: 19–25.

Ciofu O, Tolker-Nielsen T. 2019. Tolerance and resistance of *Pseudomonas aeruginosa* biofilms to antimicrobial agents: How *P. aeruginosa* can escape antibiotics? *Front Microbiol.* 10: 913. doi: 10.3389/fmicb.2019.00913.

Cirillo JD, Falkow S, Tompkins LS, Bermudez LE. 1997. Interaction of *Mycobacterium avium* with environmental amoebae enhances virulence. *Infect Immun.* 65: 3759–3767.

De Beer D, Srinivasan R, Stewart PS. 1994a. Direct measurement of chlorine penetration into biofilms during disinfection. *Appl Environ Microbiol.* 60: 4339–4344.

De Beer D, Stoodley P, Roe F, Lewandowski Z. 1994b. Effects of biofilm structures on oxygen distribution and mass transport. *Biotech Bioengin.* 43: 1131–1138.

Dey R, Rieger AM, Stephens C, Ashbolt NJ. 2019. Interactions of *Pseudomonas aeruginosa* with *Acanthamoeba polyphaga* observed by imaging flow cytometry. *Cytometry Part A.* 95A: 555–564.

Falkinham JO III. 2011. Nontuberculous mycobacteria from household plumbing of patients with nontuberculous mycobacteria disease. *Emerg Infect Dis.* 17: 419–424.

Favero MS, Carson LA, Bond WW, Petersen NJ. 1971. *Pseudomonas aeruginosa*: Growth in distilled water from hospitals. *Science.* 173: 836–838.

Furuhata K, KLoike KA, Matsumoto A. 1987. Growth and survival of a chlorine-resistive Gram-negative rod bacterium *Protomonas extorquens* isolated dominantly from drinking tank-water. *Bull Jpn Soc Microbiol Exol.* 4: 35–47.

Govan JRW, Brown AR, Jones AM. 2007. Evolving epidemiology of *Pseudomonas aeruginosa* and *Burkholderia cepacia* complex in cystic fibrosis lung infection. *Future Microbiol.* 2: 153–64. doi: 10.2217/17460913.2.2.153.2.

Grobe S, Wingender J, Flemming H-C. 2001. Capability of mucoid *Pseudomonas aeruginosa* to survive in chlorinated water. *Int J Hyg Environ Health.* 204: 139–142.

Guenette S, Williams MD, Falkinham JO III. 2020. Growth temperature, trehalose, and susceptibility to heat in *Mycobacterium avium. Pathogens.* 9: 657.

Hiraishi A, Furuhata K, Matsumoto A, Koike KA, Fukuyama M, Tabuschi K. 1995. Phenotypic and genetic diversity of chlorine-resistant *Methylobacterium* strains isolated from various environments. *Appl Environ Microbiol.* 61: 2099–2107.

Karumathil DP, Yin H-B, Kollanoor-Johny A, Venkitanarayanan K. 2014. Effect of chlorine exposure on the survival and antibiotic gene expression of multidrug resistant *Acinetobacter baumanii* in water. *Int J Environ Res Pub Hlth.* 11: 1844–1854.

Kidd TJ, Ritchie SR, Ramsay KA, Grimwood K, Bell SC, Rainey PB. 2012. *Pseudomonas aeruginosa* exhibits frequent recombination, but only a limited association between genotype and ecologic setting. *PLoS ONE.* 7: e44199. doi: 10.1371/journal.pone.0044199.

Kuchta JM, States SJ, McGlaughlin JE, Overmeyer JH, Wadowsky RM, McNamanara AM, Wolford RS, Yee RB. 1985. Enhanced chlorine resistance of tap water-adapted *Legionella pneumophila* as compared with agar medium-passaged strains. *Appl Environ Microbiol.* 50: 21–26.

Lau HY, Ashbolt NJ. 2009. The role of biofilms and protozoa in *Legionella* pathogenesis: Implications for drinking water. *Appl Microbiol.* 107: 368–378.

Lewis AH, Falkinham JO III. 2015. Microaerobic growth and anaerobic survival of *Mycobacterium avium, Mycobacterium intracellulare* and *Mycobacterium scrofulaceum. Intl J Mycobacteriol.* 4: 25–30.

Li L, Mendis N, Trigui H, Oliver JD, Faucher SP. 2014. The importance of the viable but non-culturable state in human bacterial pathogens. *Front Microbiol.* doi: 10-3389/fmicb.2014.00258.

Mullis SN, Falkinham JO III. 2013. Adherence and biofilm formation of *Mycobacterium avium, Mycobacterium intracellulare* and *Mycobacterium abscessus* to household plumbing materials. *J Appl Microbiol.* 115: 908–914.

Norton CD, LeChevallier MW, Falkinham JO III. 2004. Survival of *Mycobacterium avium* in a model distribution system. *Water Research.* 38: 1457–1466.

Palmer KL, Brown SA, Whiteley M. 2007. Membrane-bound nitrate reductase is required for anaerobic growth in cystic fibrosis sputum. *J Bacteriol.* 189: 4449–4455.

Rodriguez-Rojas A, Oliver A, Blázquez J. 2012. Intrinsic and environmental mutagenesis drive diversification and persistence of *Pseudomonas aeruginosa* in chronic lung infections. *J Infect Dis.* 205: 121–127.

Rogers J, Dowsett AB, Dennis PJ, Lee JV, Keevil CW. 1994. Influence of plumbing materials on biofilm formation and growth of *Legionella pneumophila* in potable water systems. *Appl Environ Microbiol.* 60: 1842–1851.

Ryan RP, Monchy S, Cardinale M, Taghavi S, Crossman L, Avison MB, Berg G, van der Lelie D, Dow JM. 2009. The versatility and adaptation of bacteria from the genus *Stentrophomonas. Nat Rev Microbiol.* 7: 514–525.

Sax H, Bloemberg G, Hasse B, Sommerstein R, Kohler P, Achermann Y, Rössle M, Falk V, Kuster SP, Böttger EC, Weber R. 2015. Prolonged outbreak of *Mycobacterium chimaera* infection after open-chest heart surgery. *Clin Infect Dis.* 61: 67–75.

Schofield GM, Locci R. 1985. Colonization of components of a model hot water system by *Legionella pneumophila*. *J Appl Bacteriol.* 58: 151–162.

Schulze-Röbbecke R, Buchholtz K. 1992. Heat susceptibility of aquatic mycobacteria. *Appl Environ Microbiol.* 58: 1869–1873.

Spoering AL, Lewis K. 2001. Biofilms and planktonic cells of *Pseudomonas aeruginosa* have similar resistance to killing by antimicrobials. *J Bacteriol.* 183: 6746–6751.

Steed KA, Falkinham JO III. 2006. Effect of growth in biofilms on chlorine susceptibility of *Mycobacterium avium* and *Mycobacterium intracellulare*. *Appl Environ Microbiol.* 72: 4007–4100.

Taylor RH, Falkinham JO III, Norton CD, LeChevallier MW. 2000. Chlorine-, chloramine-, chlorine dioxide- and ozone-susceptibility of *Mycobacterium avium*. *Appl Environ Microbiol.* 66: 1702–1705.

Thomas V, Herrera-Rimann K, Blanc DS, Greub G. 2006. Biodiversity of amoebae and amoebae-resisting bacteria in a hospital water network. *Appl Environ Microbiol.* 72: 2428–2438.

Wargo MJ. 2019. Is the potable water system an advantageous preinfection niche for bacteria colonizing the cystic fibrosis lung? *mBio.* 10: e00883–19.

Yano T, Kubota H, Hanai J, Hitomi J, Tokuda H. 2013. Stress tolerance of *Methylobacterium* biofilms in bathrooms. *Microbes Environ.* 28: 87–95.

Chapter 9

Detection, Isolation, and Source Tracking of OPPPs

9.1 Distinguishing between Premise Plumbing Pathogens and Contaminants

Several lines of evidence can be used to determine whether a known or unknown opportunistic premise plumbing pathogen (OPPP) is present and possibly linked to infections. These include (a) disinfectant resistance, (b) surface adherence and biofilm formation, (c) regrowth in a drinking water distribution system of premise plumbing, (d) growth at low oxygen concentrations, (e) desiccation tolerance, and (f) growth and survival in amoebae-resisting microorganisms (ARMs).

It is important to recognize that the OPPPs detected by standardized testing, such as *L. pneumophila* or *M. avium*, are not necessarily the only pathogenic members of their genera. In fact, because reporting is low or not required in most situations (except for *L. pneumophila*) and many community-acquired opportunistic pathogen infections go unidentified, the OPPPs that we are aware of are likely only a proportion of the full spectrum of OPPPs. In a practical sense, it would be of value to identify novel OPPPs to anticipate future emerging waterborne pathogens and to better inform effective plumbing design and

Opportunistic Premise Plumbing Pathogens
Joseph O. Falkinham, III

ISBN 978-981-4968-40-9 (Hardcover), 978-1-003-32100-2 (eBook)
www.jennystanford.com

engineering control of waterborne diseases. Such an objective requires guidance for the detection of emerging OPPPs.

9.2 Sources of OPPPs

Before discussing selective media for OPPPs, it is important to identify known sources for OPPPs. Although all are waterborne pathogens, it is important to consider that OPPPs prefer surface attachment where they grow on surfaces in biofilms (Rogers et al., 1994; Mullis and Falkinham, 2013). That means there are far more cells on surfaces and that one's chances of isolating an OPPP are higher starting with biofilm (swab) surface samples than bulk water. For example, numbers of *M. avium* and *P. aeruginosa* are commonly 10,000 colony-forming units (CFU)/cm^2 of pipe surface compared to 10–100 CFU/mL of water (Mullis and Falkinham, 2013). Thus, the likelihood of detecting OPPPs is higher in swab samples. Further, the effort of collecting, transporting, and analyzing a 1 L water sample is considerably higher than the analysis of a swab placed in 2 mL of water.

Table 9.1 Sources of OPPPs

Drinking Water
Showerheads
Water Taps
Water Tap Aerators
Ice and Water from Refrigerators
Granular Activated Carbon (Charcoal) Filters (GAC)
Water Heaters
Humidifiers and Humidifier Aerosols
Humidifiers in Heating Ventilation Air Conditioning Units

In Table 9.1, premise plumbing sources are listed that have been identified as sources for OPPPs. We were led to one source, refrigerator water, as a result of the diligence of the son of a *Mycobacterium* spp. infected patient based on the son's analysis of possible water sources and his parent's water exposures. As is the case in medical schools where future physicians are taught the importance of getting a case history, the same lesson holds here for identifying the source of an OPPP infection. Ask about the patient's activities.

9.3 Selective Media for Recognized OPPPs

Table 9.2 lists emerged and emerging OPPPs, agar media suitable for their selective isolation, and references. All are based on unique characteristics of the microorganisms and, in a majority of cases, the selective isolation media have been tested.

Table 9.2 Selective media for isolation of OPPPs

OPPP	Medium Name	Reference(s)
L. pneumophila	Edelstein, 1982	Ducret et al., 2014 CDC, 2005
M. avium	Tsukamura Tween 80	George & Falkinham, 1986
P. aeruginosa	*Pseudomonas*	Lowbury & Collins, 1955
	Isolation Agar	Goto & Enomoto, 1970
A. baumannii	LEEDS Medium	Jawad et al., 1994
S. maltophilia	Met + Imipenem	Bollet et al., 1995
S. paucimobilis	Yellow, PEG Agar	Takeuchi et al., 1993
	3% NaCl @ 40 °C	Yabuuchi et al., 1990
Methylobacterium	Pink Methanol Agar	Green & Ardley, 2018
	Glyoxylate Agar	Takeuchi et al., 1993
Segniliparus	M7H10 + Malachite	Butler et al., 2005
Cupriavidus	000 µM Cu^{+2} + 0.1% L-alanine Agar	Makkar & Casida, 1987
Acanthamoeba spp.	R2A + *E. coli* Agar	

9.4 The Question of Cultivation versus DNA-Based Detection

As the majority of our sampling is aimed at identifying the source of an OPPP infection, we culture samples. That allows us to fingerprint isolates and archive frozen cells and DNA for further study. Yes, it takes more time, especially for slow-growing OPPPs. This direct plating approach suffers from a lack of detection if the potential OPPP is prone to enter a viable, but non-culturable (VVNC) state.

9.5 The Problem of VBNC State

At the outset of this discussion, it should be noted that the emerged (Chapter 4) and emerging new OPPPs (Chapter 7) can be in a VBNC state, and detection by colony formation might fail (Li et al., 2014). VBNC cells display respiratory activity (e.g., redox dye reduction), membrane integrity (e.g., membrane permeation by dyes), yet can be counted as cells using fluorescent in situ hybridization (FISH) with species-specific 16S or 18S rRNA probes or total cells counts (Li et al., 2014; Ducret et al., 2014). The VBNC state is triggered by exposure to stress associated with exposure to: disinfectants, reduced temperature, toxic oxygen metabolites, or starvation (Li et al., 2014). During sampling and isolation protocols, consideration should be given to providing "recovery" conditions for cells in the VBNC state, before proceeding with detection based on colony formation on agar media (Ducret et al., 2014). "Recovery" conditions include incubation in broth media, incubating samples or cells suspensions in toxic oxygen metabolite scavengers (e.g., pyruvate or mannitol), or even incubation in amoebae (Thomas et al., 2014; 2008). Employing the approach of the first isolating amoebae and then ARMs, *Methylobacteria*, mycobacteria, and *Legionella* were recovered from a hospital water network (Ducret et al., 2014). In another study, amoebae in water yielded: *A. baumannii, A. hydrophila, L. pneumophila, Methylobacterium mesophilicum, M. avium, P. aeruginosa*, and *Stenotrophomonas maltophilia* (Cateau et al., 2011; 2014).

9.6 Detection and Isolation of Novel OPPPs

Toward the objective of detecting novel and emerging OPPPs, the following methods for isolation of putative OPPPs are offered. The approaches are based on the common, shared characteristics of OPPPs (Chapter 6). It is probable that a substantial proportion of microbes isolated will not be pathogenic. However, they will share some characteristics in common with OPPPs (e.g., amoebae resisting and biofilm formation) and in as much the number of individuals more susceptible to microbial infections is increasing, it is prudent to be alert to emerging OPPPs.

9.6.1 Selection for ARMs

One of the mechanisms contributing to the survival and persistence of waterborne bacteria in drinking water distribution systems and premise plumbing is the shelter from disinfectant provided by phagocytic amoebae. The most direct way for isolating and identifying such bacterial species (and candidate OPPPs) is to recover amoebae and protozoa from drinking water or biofilm samples (Ovrutsky et al., 2013; Thomas et al., 2008). Those methods involve the collection of amoebae by low-speed centrifugation (1,000 × g), washing the pelleted amoebae free of suspended microbes in the supernatant liquid. Subsequent lysis of the washed amoebae with low concentrations of detergent (e.g., 0.1% Tween 80 or Triton X-100) and spread plating on a medium suitable for growth should yield colonies resulting from any intracellular bacteria. Fortunately, most OPPPs are relatively resistant to the low detergent concentrations employed to lyse the amoebae. As many microbial dyes will stain cells within amoebae (e.g., the Gram-stain or the acid-fast stain), amoeba concentrates can be rapidly screened to determine whether they harbor intracellular bacteria before lysis. This approach has been systematically described in two publications (Thomas et al., 2008; Ovrutsky et al., 2013).

9.6.2 Selection for Disinfectant-Resistant Microorganisms

One characteristic shared by OPPPs is their resistance to disinfectants (e.g., chlorine and chloramine) used in water treatment. As the water in some distribution systems and premise plumbing has a residual disinfectant concentration, any surviving microbe in those habitats must be disinfectant-resistant. Amoebae can enhance the resistance of resident bacteria, particularly when amoebae are in the cyst form, and protect cells that are not inherently disinfectant-resistant (e.g., *Legionella* spp.; Thomas et al., 2004). The disinfectant residual cannot only select resistant cells, but also diminish disinfectant-sensitive competitors, allowing putative OPPPs to proliferate utilizing available organic carbon. Thus, disinfectant resistance can also be used as a selective measure to isolate OPPPs. Microorganisms can be directly selected by adding a disinfectant to a water or biofilm sample at a concentration high enough to kill *E. coli*, but not so high that all microbes are killed (e.g., 1 ppm for 30, 60, and 120 min).

9.6.3 Selection of Biofilm-Forming Microorganisms

Another shared characteristic of OPPPs is their tendency to adhere to surfaces and proliferate in biofilms. Biofilm formation is essential for the persistence and survival of a microorganism in a flowing pipe system. This is particularly the case for slow-growing opportunistic pathogens such as the mycobacteria (Mullis and Falkinham, 2013) and *Legionella* (Rogers et al., 1994), and their growth rates are not high enough to replace cells lost due to liquid flow. Biofilm formation, signaled by aggregation of cells in culture (Green and Ardley, 2018), may also be a corollary of amoebae resistance, as amoebae phagocytosis is an important factor governing the population composition of biofilms. Coupons that are composed of pipe material (e.g., copper, PVC, or stainless steel) can be placed in a water sample for 2–6 h at room temperature and adherent microorganisms can be scraped off or suspended by rapid mixing after gentle washing to suspend cells. In this approach, the pipe material coupon acts as a selective agent to pull adherent microorganisms out of suspension. Different surface materials (coupons) can be placed in water, as a variety of pipe materials (e.g., glass, PVC, stainless steel, copper, and galvanized steel) are used in premise plumbing, and microorganisms display different hierarchies of adherence. For example, *M. avium* adherence differs significantly between pipe materials; adherence is the highest to galvanized steel, and the lowest to glass (Mullis and Falkinham, 2013).

In a study of adherence of *L. pneumophila* and *P. aeruginosa* to pipe materials used in premise (domestic) plumbing, it was shown that *P. aeruginosa* failed to adhere to copper surfaces but *L. pneumophila* did (Bédard et al., 2016). The authors ascribed this observation to the reported sensitivity of *P. aeruginosa* to copper ions (Bédard et al., 2016). Thus, it is important to select a surface where one suspects the opportunistic pathogen is in the biofilm.

9.6.4 Heat Resistance

As a number of the established and emerging OPPPs have been shown to tolerate exposure to high temperature, namely *L. pneumophila* (Whiley et al., 2017), *M. avium* complex (Guenette et al., 2020), samples could be exposed to elevated temperatures (e.g., greater than 50 °C for 5 min) and survivors plated on selective and non-selective media.

9.6.5 Growth at Low Oxygen

As OPPPs must be able to survive during desiccation, a water or biofilm sample can be placed in a chamber, oxygen concentration lowered by various means (e.g., even a candle jar approach), and following a period of incubation, samples plated for colonies.

9.6.6 Desiccation Resistance

Desiccation resistance is a characteristic of *Methylobacterium* spp. (Yano et al., 2013) and *Mycobacterium* spp. (Guenette et al., 2020), and samples could be concentrated and 0.1 mL added to sterile filter paper and incubated at low relative humidity (e.g., desiccant) for 1–3 weeks (Hugenholtz et al., 1995). Alternatively, sample concentrates could be spotted on surfaces (e.g., coupons) of materials commonly used in premise plumbing (e.g., stainless steel, galvanized, or PVC) and incubated for up to 1–3 months. Survivors, as above, can be plated on selective or non-selective media for detection, isolation, and enumeration.

9.7 Source Tracking

9.7.1 Sample Selection

First, ask the patients or hospital workers to show where they work and live; namely, identify possible sources of exposure and infection. With that approach, you will sample unconventional sites as noted above in identifying a refrigerator water tap as *Mycobacterium* spp. source. Sample potting soil and soil in pots and gardens if an individual is a gardener. We have been successful in identifying that potting soil was the source of *M. intracellulare* infection in a number of gardeners (de Groote et al., 2006).

Although all of the OPPPs I have included here are waterborne (required), rather than collect water samples, I suggest the collection of biofilm (swab) samples. First, in my experience, there are many more cells in biofilms than floating in suspension. Second, if the OPPP is in the water, it will be in the water tap or showerhead; both of which can be sampled with a swab. I swab a 2 cm × 2 cm area and place the swab in 2 mL of sterile tap water (not distilled water). By the time, the swab is returned to my lab (2–7 days), the microbial cells have been suspended in the water and with a short vortex, 0.1 mL of the swab's suspended cells can be spread on the selective or non-selective medium agar. You can always collect water samples, but that might involve shipping 500–1,000 mL of water. We always concentrate a water sample down to 2 mL.

As with any cultivation approach, OPPP colonies can often be overgrown by fast-growing microorganisms. This can be a problem when trying to isolate pseudomonads or mycobacteria from biofilm or soil samples; every other microorganism can grow faster and obscure mycobacterial colonies. Water concentrates, biofilm suspensions, or soil suspensions are routinely treated with 0.01% cetylpyridinium chloride (CPC) for 30 min to reduce the number of non-mycobacterial cells (Williams and Falkinham, 2018). Although at that dosage mycobacterial colony numbers are reduced (10%), almost every rapidly growing microorganism is killed (> 99.9%; Williams and Falkinham, 2018). One must remember that because CPC reduces the number of mycobacteria, the number of colonies is always less than the actual number if CPC-decontamination is employed. The use of CPC has also proven to be effective in isolating other OPPPs; specifically *Acinetobacter baumannii* and *Stenotrophomonas maltophilia* (Falkinham, unpublished), the only modification is to reduce the dosage of CPC, as other OPPPs are not as resistant (2- to 10-fold) to CPC than are the mycobacteria.

9.7.2 Fingerprinting

Isolation of cultures allows us to fingerprint isolates from patients and their environment to identify likely sources of infection. When initiating a source identification study, remember that spontaneous mutation occurs in both the patient and environment. In fact, there are a number of papers demonstrating changes in DNA fingerprints or whole-genome sequences of OPPPs in either patients or their environments, or both. With *M. avium*, we find that the patient and their environmental isolates belong to the same clone; they are not always identical (Falkinham et al., 2008).

There are a variety of methods for fingerprinting OPPP isolates. For *M. avium*, we have used pulsed field gel electrophoresis (PFGE) to compare band patterns of large chromosomal fragments (von Reyn et al., 1994). Unfortunately, PFGE analysis is laborious and time-consuming. We have also used insertion sequence (IS) typing that relies on the random transposition of IS in the genome (Falkinham et al., 2008). DNA is isolated from different strains and digested with restriction endonucleases to generate a family of fragments. Using a specific tagged IS element as a probe, the DNA fragments carrying the IS can be identified and similarity judged by the number of shared fragments carrying the IS. This approach is also time-consuming and requires multiple, separate reactions. My favorite is *rep*-PCR in which sequences between repetitive elements (*rep*) are amplified by PCR (Falkinham, 2011). If strains are identical, the number of PCR amplicons and their size will be the same. As *rep*-PCR requires a single PCR reaction followed by gel electrophoresis, it is simple and rapid. Repetitive genetic elements are present in almost every bacterial species, thereby providing one with the tool to use *rep*-PCR. Recently, it has been shown that tandem repeats of sequences can be used for fingerprinting (Iakhiaeva et al., 2016). In the mycobacteria, mycobacterial internal repetitive unit—variable number tandem repeats (MIRU-VNTR) can be used for fingerprinting. In the *M. avium* complex, different sequence tandem repeats have been identified and the number of tandem copies of the repeat sequence identified to judge the variation of each.

For all these fingerprinting techniques, one must determine the discrimination index (Hunter and Gaston, 1988) for each. Specifically, for fingerprinting one needs to have a method that will generate a large number of DNA fragments that are detected, whether they be large genome fragments (PFGE), IS-linked fragments, *rep*-PCR amplicons, or tandem repeats (VNTR). measured. If the number of countable fragments is low, the index of discrimination is too low to provide certainty of origin.

Finally, whole-genome sequencing has been employed for source tracking (Lande et al., 2019). Here, rather than spontaneous changes in a restriction endonuclease site, movement of an insertion (IS) or repetitive (*rep*) sequence, or the number of tandem repeats (VNTR), all changes in the genome can be detected. However, it is important to have a standard for comparison involving strains of different origins. Not just for whole-genome sequence analysis, it is wise to obtain serial isolates of a single species infecting a patient or resident in premise plumbing and compare variation over time. With that knowledge, one can judge the significance of differences between isolates. Such an analysis of consecutive, serial isolates of *P. aeruginosa* recovered from cystic fibrosis patients (Bianconi et al., 2019).

References

Bédard E, Prévost M, Déziel E. 2016. *Pseudomonas aeruginosa* in premise plumbing of large buildings. *Microbiol Open*. 5: 937–956.

Bianconi I, D'Arcangelo S, Esposito A, Benedet M. Fiffer E, Dinnella G, Gualdi P, Schinella M, Baldo E, Donati C, Jousson O. 2019. Persistence and microevolution of *Pseudomonas aeruginosa* in the cystic fibrosis lung: A single-patient longitudinal genomic study. *Front Microbiol*. 9: 3242.

Bollet C, Davin-Regli A, De Micco P. 1995. A simple method for selective isolation of *Stenotrophomonas maltophilia* from environmental samples. *Appl Environ Microbiol.* 61: 1653–1654.

Butler WR, Floyd MM, Brown JM, Toney SR, Daneshvar MI, Cooksey RC, Carr J, Steigerwalt AG, Charles N. 2005. Novel mycolic acid-containing bacteria in the family *Segniliparaceae* fam. nov., including the genus *Segniliparus* gen. nov., with descriptions of *Segniliparus rotundus* sp. nov. and *Segniliparus rugosus* sp. nov. *Int J System Evol Microbiol.* 55: 1615–1624.

Cateau E, Maisonneuve E, Peguilhan S, Quellard N, Hechard Y, Rodier M-H. 2014. *Stenotrophomonas maltophilia* and *Vermamoeba vermiformis* relationships: Bacterial multiplication and protection in amoebal-derived structures. *Res Microbiol.* XX: 1–5.

Cateau E, Verdon J Fernandez B, Hechard Y, Rodier M-H. 2011. *Acanthamoeba* spp. promotes the survival and growth of *Acinetobacter baumannii*. *FEMS Microbiol Lttrs.* 319: 19–25.

Centers for Disease Control. 2005. Procedures for the recovery of *Legionella* from the environment. https://www.cdc.gov/legionella/labs/procedures-manual.html.

De Groot MA, Pace NR, Fulton K, Falkinham JO III. 2006. Relationships between *Mycobacterium* isolates from patients with pulmonary mycobacterial infection and potting soils. *Appl Environ Microbiol.* 72: 7062–7606.

Ducret A, Cabalier M, Dukan S. 2014. Characterization and resuscitation of 'non-culturable' cells of *Legionella pneumophila*. *Microbiol.* 14: 3. www.biomedcentral.com/1471-2180/14/3.

Edelstein PH. 1982. Comparative study of selective media for isolation of *Legionella pneumophila* from potable water. *J Clin Microbiol.* 16: 697–699.

Falkinham JO III. 2011. Nontuberculous mycobacteria from household plumbing of patients with nontuberculous mycobacteria disease. *Emerg Infect Dis.* 17: 419–424.

Falkinham JO III, Iseman MD, de Haas P, van Soolingen D. 2008. *Mycobacterium avium* in a shower linked to pulmonary disease. *J Water Health.* 6: 209–213.

George KL, Falkinham JO III. 1986. Selective medium for the isolation and enumeration of *Mycobacterium avium-intracellullare* and *M. scrofulaceum*. *Canad J Microbiol.* 32: 10–14.

Goto S, Enomoto S. 1970. Nalidixic Acid Cetrimide Agar. A new selective plating medium for the selective isolation of *Pseudomonas aeruginosa*. *Japan J Microbiol.* 14: 65–72.

Green PN, Ardley JK. 2018. Review of the genus *Methylobacterium* and closely related organisms: A proposal that some *Methylobacterium* species be reclassified into a new genus, *Methylorubrum* gen. nov. *Int J Syst Evol Microbiol.* 68: 2727–2748.

Guenette S, Williams MD, Falkinham JO III. 2020. Growth temperature, trehalose, and susceptibility to heat in *Mycobacterium avium*. *Pathogens.* 9: 657. doi: 10.3390/pathogens9080657.

Hugenholtz P, Cunningham MA, Hendrlkz JK, Fuerst JA. 1995. Desiccation resistance of bacteria isolated from an air-handling system biofilm determined using a simple quantitative membrane filter method. *Lttrs. Appl. Microbiol.* 21: 41–46.

Hunter PR, Gaston MA. 1988. Numerical index of the discriminatory ability of typing systems: An application of Simpson's index of diversity. *J Clin Microbiol.* 26: 2465–2466.

Iakhiaeva E, Howard ST, Brown-Elliott BA, McNulty S, Newman KL, Falkinham JO III, Williams M, Kwait R, Lande L, Vasireddy R, Turenne C, Wallace RJ Jr. 2016. Variable number tandem repeat (VNTR) of respiratory and household water biofilm isolates of *Mycobacterium avium* subspecies "*hominissuis*" with establishment of a PCR database. *J Clin Microbiol.* 54: 891–901.

Jawad A, Hawkey PM, Heritage J, Snelling AM. 1994. Description of Leeds *Acinetobacter* medium, a new selective and differential medium for isolation of clinically important *Acinetobacter* spp., and comparison with Herellea agar and Holton's agar. *J Clin Microbiol.* 32: 2353–2358.

Lande L, Alexander DC, Wallace RJ Jr, Kwait R, Iakhiaeva E, Williams M, Cameron ADS, Olshefsky S, Devon R, Vasireddy R, Peterson DD, Falkinham JO III. 2019. *Mycobacterium avium* in community and household water, suburban Philadelphia, Pennsylvania, USA, 2010–2012. *Emerg Infect Dis.* 25: 473–481.

Li L, Mendis N, Trigui H, Oliver JD, Faucher SP. 2014. The importance of the viable, but non-culturable state in human bacterial pathogens. *Front Microbiol.* doi: 10.3389/fmicb.2014.00258.

Lowbury JL, Collins AG. 1955. The use of a new cetrimide product in a selective medium for *Pseudomonas pyocyanea*. *J Clin Path.* 8: 47–48.

Makkar NS, Casida LE Jr. 1987. *Cupriavidus necator* gen. nov., sp. nov.: A nonobligate bacterial predator of bacteria in soil. *Int J System Bacteriol.* 37: 323–326.

Mullis SN, Falkinham JO III. 2013. Adherence and biofilm formation of *Mycobacterium avium, Mycobacterium intracellulare,* and *Mycobacterium abscessus* to household plumbing materials. *J App Microbiol.* 115: 908–914.

Ovrutsky AR, Chan ED, Kartalija M, Bai X, Jackson M, Gibbs S, Falkinham JO III, Isman M, Reynolds G, McDonnell G., Thomas V. 2013. Co-occurrence of free-living amoebae and non-tuberculous mycobacteria in hospital water networks, and preferential growth of *Mycobacterium avium* in *Acanthamoeba lenticular. App Environ Microbiol.* 79: 3185–3192.

Rogers J, Dowsett AB, Dennis PJ, Lee JV, Keevil CW. 1994. Influence of plumbing materials on biofilm formation and growth of *Legionella pneumophila* in potable water systems. *Appl Environ Microbiol.* 60: 1842–1851.

Takeuchi M, Kawai F, Shimada Y, Yokota A. 1993. Taxonomic study of polyethylene glycol-utilizing bacteria: Emended description of the genus *Sphingomonas* and new descriptions of *Sphingomonas macrogoltabidus* sp. nov., *Sphingomonas sanguis* sp. nov. and *Sphingomonas terrae* sp. nov. *System Appl Microbiol.* 16: 227–238.

Taylor RH, Falkinham JO III, Norton CD, LeChevallier MW. 2000. Chlorine, chloramine, chlorine dioxide, and ozone susceptibility of *Mycobacterium avium. Appl Environ Microbiol.* 66: 1702–1705.

Thomas V, Bouchez T, Nocolas V, Robert S, Loret JF, Lévi Y. 2004. Amoebae in domestic water systems: Resistance to disinfection treatments and implication in *Legionella* persistence. *J Appl Microbiol.* 97: 950–963.

Thomas V, Loret J-F, Greub G. 2008. Biodiversity of amoebae and amoebae-resisting bacteria in drinking water treatment plant. *Environ Microbiol.* 10: 2728–2745.

von Reyn CF, Maslow JN, Barber TW, Falkinham JO III, Arbeit RD. 1994. Persistent colonization of potable water as a source of *Mycobacterium avium* infection in AIDS. *Lancet.* 343: 1137–1141.

Whiley H, Bentham R, Brown MH. 2017. *Legionella* persistence in manufactured water systems: Pasteurization potentially selecting for thermal tolerance. *Front Microbiol.* 8: 1330. doi: 10.3389/fmicb.2017.01330.

Williams MD, Falkinham JO III. 2018. Effect of cetylpyridinium chloride (CPC) on colony formation of common nontuberculous mycobacteria. *Pathogens.* 7: 79. doi:10.3390/pathogens7040079.

Yabuuchi E, Yano I, Oyaizu H, Hashimoto Y, Ezaki T, Yamamoto H. 1990. Proposals of *Sphingomonas paucimobilis* gen. nov. and comb. nov., *Sphingomonas parapaucimobilis* sp. nov., *Sphingomonas yanoikuyae* sp. nov., *Sphingomonas adhaesiva* sp. nov., *Sphingomonas capsulata* com. nov., and two genospecies of the genus *Sphingomonas. Microbiol Immunol.* 34: 99–119.

Yano T, Kubota H, Hanai J, Hitomi J, Tokuda H. 2013. Stress tolerance of *Methylobacterium* biofilms in bathrooms. *Microbes Environ.* 28: 87–95.

Chapter 10

OPPP Notification and Challenges to Current Water Treatment Practices

10.1 Introduction

This brief chapter is included to introduce the pros and cons of requiring notification of opportunistic premise plumbing pathogen (OPPP)-disease and to list and describe the challenges to current water treatment practices by the OPPPs. As the drinking water industry is sensitive and responsive to the emergence of pathogens—witness the reactions to the COVID-19 pandemic—it is of value to identify the challenges to permit stakeholders the opportunity to devise changes in current practice to reduce the impact of OPPPs on public health and the water industry. Remember, the economic impact of *L. pneumophila, P. aeruginosa*, and *M. avium* complex waterborne infections and diseases is huge; approximately over $1 billion annually (Collins et al., 2012).

10.2 Notification

What are the possible benefits of requiring notification of OPPP infections and diseases? To date, only infections caused by *L. pneumophila* must be reported and almost 12 states now require notification of *M. avium* complex infections. The benefits

Opportunistic Premise Plumbing Pathogens
Joseph O. Falkinham, III

ISBN 978-981-4968-40-9 (Hardcover), 978-1-003-32100-2 (eBook)
www.jennystanford.com

of notification include (a) detection of clusters and outbreaks, (b) planning public health interventions, and (c) evaluation of public health interventions. Using *L. pneumophila* as an example, notification leads to rapid recognition of *L. pneumophila* outbreaks and public health authorities can mount searches linking infections to point sources, and institute remediation measures. The notification provides the data to calculate the prevalence and incidence of Legionnaires' disease from year-to-year and identify trends that will, in turn, permit better planning for the public health responses needed to either combat newly emerging outbreaks or plan for future responses. Because of notification, the efficacy of remediation measures can be measured and judged. Currently, outbreaks of OPPP infections in hospitals can be investigated to identify the sources and institute and measure the effectiveness of remediation measures.

Notification may also require widespread implementation of expensive protocols to monitor OPPPs in the human environment. As OPPPs are in drinking water and thereby everywhere, what sites and samples should be monitored? Further, in the absence of knowledge of infective dosages for OPPPs, save a few (Rusin et al., 1997; Hamilton et al., 2017), what numbers would require action?

10.3 Challenges to Current Water Treatment Practices

10.3.1 OPPP Numbers Increase from a Point Source

Microbial water quality is generally evaluated by the measurement of numbers of *Escherichia coli* or fecal coliforms. These evolved out of the discovery that those were indicators of fecal contamination of water supplies and further, they were surrogates or indicators for pathogenic contaminants of fecal origin such as *Salmonella* spp., *Shigella* spp., and *Vibrio* spp. Thus, evidence of the presence of *E. coli* or fecal coliforms was evidence of fecal contamination of water.

E. coli, fecal coliforms, *Salmonella* spp., *Shigella* spp., *Vibrio* spp., or *Cryptosporidium* spp. do not grow in drinking water but are killed following entry. Therefore, numbers of any of those water microbial contaminants fall as the distance from the point of fecal contamination increases. To identify the source of fecal contamination, one can measure the numbers of *E. coli* or fecal coliforms moving against the flow in a stream or a river or at different sites in a pond or a lake. Numbers are the highest at the point of entry. That is where action should be taken to prevent further contamination.

In contrast, OPPPs are adapted and grow in drinking water distribution systems; they are colonists. For example, *M. avium n*umbers increase with distance from a treatment plant (Falkinham et al., 2001). OPPPs are not simply transported through a distribution system, but grow during their movement. So source tracking by following an increase in numbers in a flowing system would not identify the source; in fact, the source is the water itself.

10.3.2 What Samples to Collect for OPPP Monitoring?

It is not clear what type of samples needs to be collected for OPPP monitoring. OPPP numbers increase during the transport from a treatment plant to a water meter and then within premise plumbing. Where should the samples be collected? What type of samples should be collected: bulk water or biofilms? As it is likely that many OPPP infections occur in households, the variation in household practices, such as water heater temperature, the presence/absence of faucet aerators, and a possible stagnation, will directly influence OPPP infection probabilities.

10.3.3 Unknown Dose-Response Values

As noted above in the discussion of notification, dose-response values for OPPP infections are unknown. In the absence of established dose-response values, numerical values of OPPPs in drinking water bulk or biofilm samples cannot be compared to an established level requiring remediation. Further, as the OPPPs are opportunistic pathogens, humans, even those with risk factors, susceptibility can vary over a wide range. OPPP numbers can be obtained, but their public health impact is unknown.

10.3.4 OPPP Numbers Do Not Correlate with *E. coli*, Fecal Coliforms, and Heterotrophic Plate Count (HPC)

A variety of studies have shown that *E. coli* and fecal coliform counts or presence/absence do not correlate with OPPP numbers (Falkinham et al., 2001; Briancesco et al., 2014; Hamilton et al., 2016). This is not surprising as the OPPPs are natural inhabitants of water—colonists, not contaminants— and capable of survival, growth, and persistence in premise plumbing.

Another measure of the microbiological quality of water, the HPC is often employed to assess the microbial quality of drinking water. However, HPC values do not correlate with OPPP numbers (Allen et al., 2004; Bartram et al., 2004). Further, OPPP numbers cannot be extracted from total HPC numbers. Choice of the HPC is logical over *E. coli* and fecal coliform counts, as the OPPPs are heterotrophic bacteria. A standard for microbial quality of drinking water is that the HPC should be below 500 colony-forming units (CFU)/mL. A number of studies have shown that there is no evidence that the HPC values "...support health-based regulations..." (Allen et al., 2004; Bartram et al., 2004). There are a number of reasons for this lack of correlation. First, there is no uniform, accepted single method for obtaining HPC values (Bartram et al., 2004; Gensberger et al., 2015). Second, although members of the genus *Mycobacterium* grow on the commonly used HPC medium, R2A, the incubation period is too short (2 days) to allow for the appearance of any *Mycobacterium* spp. colonies. The foregoing is not to argue against the use of HPC values for other purposes; simply, it is not useful for providing any guidance concerning the presence/absence or numbers of OPPPs.

10.3.5 Is *Pseudomonas aeruginosa* a Surrogate or Indicator for OPPPs?

In the absence of any indicator or surrogate for OPPPs, it might be possible for an OPPP to be employed as a surrogate; for example, *P. aeruginosa*. First, *P. aeruginosa* shares many features, if not all, with other OPPPs (Chapter 4.2). One attractive feature of *P. aeruginosa* is its rapid growth and availability of selective media permitting simultaneous exclusion of other waterborne microorganisms and direct identification (Chapter 9, Table 9.2). Specifically, *P. aeruginosa* forms distinctive blue-green pigmented colonies with a yellow fluorescence after 48 h incubation at 37 °C in 48 h on *Pseudomonas* isolation agar (Sartory et al., 2015).

P. aeruginosa presence was shown to be an indicator of human health risk in a study of drinking water samples in Mexico (de Victorica and Galván, 2001; Loveday et al., 2014). In studies of the microbial flora of heater-coolers, we discovered that there was a correlation between *P. aeruginosa* and *Mycobacterium chimaera* numbers (Falkinham, 2017). Although these few studies cannot offer proof of the utility of *P. aeruginosa* as an indicator of OPPPs, they do offer a direction for further study.

10.3.6 Disinfection Following *E. coli* Guidance Selects for OPPPs

Water disinfection guidance is based on the susceptibility of *E. coli* to disinfectants; commonly through the use of the metric $EC_{99.9\%}$, the product of the concentration of disinfectant (mg/mL or ppm), and exposure in minutes to kill 99.9% of cells. Exposure duration of chlorine for killing 99.9% of *E. coli*, at 1 mg/mL is 3 s ($EC_{99.9\%}$ = 0.05). Unfortunately, that dosage is way too low to kill *M. avium* ($EC_{99.9\%}$ = > 100, Taylor et al., 2000) or *P. aeruginosa* ($EC_{99.9\%}$ = 15, Bédard et al., 2016). To kill 99.9% of *M. avium* cells with 1 mg chlorine/mL would require an exposure duration of 3 h. Thus, a protocol involving exposure to a disinfectant able to kill 99.9% of *E. coli* cells will have little effect on *M. avium* or *P. aeruginosa* numbers. That is responsible, in part, for the fact that disinfectant treatment of drinking water selects rather than kills *M. avium* and *P. aeruginosa*.

The use of employing data for *E. coli* to guide disinfection of OPPPs in drinking water is further exacerbated by the preference of OPPPs to form adhere to premise plumbing pipe surfaces and form biofilms. Microbial cells in biofilms are more tolerant of disinfectants as they are unable to penetrate through the extracellular matrix and layers of cells (De Beer et al., 1994).

Depending upon the OPPP species, the equilibrium between surface adherence and suspension can be quite extreme. For example, inoculation of *M. chimaera* cells into the water of a heater-cooler leads to the loss of 99.9% of cells due to their adherence to the instrument's interior surfaces (Falkinham, 2020).

References

Allen MJ, Edberg SC, Reasoner DJ. 2004. Heterotrophic plate count bacteria: What is their significance in drinking water? *Int J Food Microbiol.* 92: 265–274.

Bartram J, Cotruvo J, Exner M, Fricker C, Glasmacher A. 2004. Heterotrophic plate count measurement in drinking water safety management. Report of an expert meeting Geneva, 24–25 April 2002. *Int J Food Microbiol.* 92: 241–247.

Bédard E, Prévost M, Déziel E. 2016. *Pseudomonas aeruginosa* in premise plumbing of large buildings. *Microbiol Open.* 5: 937–956.

Briancesco R, Alaimo C, Bonanni E, Delle Site A, Di Gianfilippo F, Grassano L, Moscatelli R, Ottaviano C, Paradiso R, Qunitiliani S, Semproni M, Bonadonna L. 2014. An Italian investigation on non-tuberculous mycobacteria in an urban water supply. *Ann Ig.* 26: 264–271. doi 10.7416/ai.20141984

Collier SA, Stockman LJ, Hicks, LA, Garrison LE, Zhou FJ, Beach, MJ. 2012. Direct healthcare costs of selected diseases primarily or partially transmitted by water. *Epidemiol Infect.* 140: 2003–2013. doi:10.1017/S0950268811002858.

De Beer D, Srinivasan R, Stewart PS. 1994. Direct measurement of chlorine penetration into biofilms during disinfection. *Appl Environ Microbiol.* 60: 4339–4344.

de Victorica J, Galvin M. 2001. *Pseudomonas aeruginosa* as an indicator of health risk for human consumption. *Water Sci Technol.* 43: 49–52.

Dias VCF, Durand A-A, Constant P, Prévost M, Bédard E. 2019. Identification of actors affecting bacterial abundance and community structures in a full-scale chlorinated drinking water distribution system. *Water.* 11: 27. doi: 10.3390/w11030627.

Falkinham JO III. 2017. *Pseudomonas aeruginosa* as an indicator for opportunistic premise plumbing pathogens. *JSM Microbiol.* 5: 1042.

Falkinham JO III. 2020. Disinfection and cleaning of heater-cooler units: Suspension- and biofilm-killing. *J Hosp Infect.* 105: 552–557.

Falkinham JO III, Norton CD, LeChevallier MW. 2001. Factors influencing numbers of *Mycobacterium avium, Mycobacterium intracellulare,* and other mycobacteria in drinking water distribution systems. *Appl Environ Microbiol.* 67: 1225–1231.

Gensberger ET, Gössl E-M, Antonelli L, Sessitsch A, Kostić T. 2015. Effect of different heterotrophic plate count methods on the estimation of the composition of the culturable microbial population. *Peer J.* 3: e862. doi 10.7717;peerj.862.

Hamilton KA, Weir MH, Haas CN. 2017. Dose-response models and quantitative microbial risk assessment framework for the *Mycobacterium avium* complex that account for recent developments in molecular biology, taxonomy, and epidemiology. *Water Res.* 109: 310–326.

Loveday HP, Wilson JA, Kerr K, Pitchers R, Walker JT, Browne J. 2014. Association between healthcare water systems and *Pseudomonas aeruginosa* infections: A rapid systematic review. *J Hosp Infect.* 86: 7–15.

Rusin PA, Rose JB, Haas CN, Gerba CP. 1997. Risk assessment of opportunistic bacterial pathogens in drinking water. *Rev Environ Contam Toxicol.* 152: 57–83.

Sartory DP, Pauly D, Garrec N, Bonadonna L, Semproni M, Schell C, Reimann A, Firth SJ, Thom C, Hartemann P, Exner M, Baldauf H, Lee S, Lee JV. 2015. Evaluation of an MPN test for the rapid enumeration of *Pseudomonas aeruginosa* in hospital waters. *J Water Health.* 13: 427–435.

Taylor RM, Norton CD, LeChevallier MC, Falkinham JO III. 2000. Chlorine-, chloramine-, chlorine dioxide- and ozone-susceptibility of *Mycobacterium avium. Appl Environ Microbiol.* 66: 1702–1705.

Chapter 11

Management and Remediation of OPPPs

11.1 Introduction

Taken together, a review of the characteristics of opportunistic premise plumbing pathogens (OPPPs) (Chapter 6) suggests that they are ideally suited for growth in premise plumbing. Further, those shared characteristics suggest that susceptibility to OPPP reduction measures demonstrated by one OPPP might be shared by all. It thereby follows that for remediation efforts to be successful in reducing OPPP numbers, the premise plumbing environment must be changed. OPPPs survive disinfection and are not washed out of pipes because of biofilm formation. Phagocytic, free-living amoebae are also not going to reduce OPPP numbers as either the bacteria grow within the amoebae (i.e., *L. pneumophila* and *M. avium*) or kill the amoebae (i.e., *P. aeruginosa*). In the absence of competitors, OPPPs can utilize even the low concentrations of organic carbon to grow, perhaps not even restricted by low oxygen concentrations characteristic of stagnation.

One way to address concerns that OPPPs may have colonized a drinking water system and everything coming in contact with it is through hazard analysis and critical control point (HACCP) analysis. HACCP involves the following seven steps: (a) hazard analysis, (b) critical control point identification, (c) establishment

Opportunistic Premise Plumbing Pathogens
Joseph O. Falkinham, III

ISBN 978-981-4968-40-9 (Hardcover), 978-1-003-32100-2 (eBook)
www.jennystanford.com

of critical limits, (d) identifying monitoring procedures, (e) making corrective actions, (f) maintaining records, and (g) verification procedures. Although commonly applied to office buildings and hospitals, those same steps could be performed by a homeowner with risk factors for OPPP disease.

11.2 HACCP Analysis of Hospital *Mycobacterium chimaera* Outbreaks

One example of hospital-level HACCP analysis can be taken from the evidence of a cluster of *M. chimaera* infections in patients following cardiac surgery; in particular heart transplant patients (Sax et al., 2015). Such infections occurred throughout the world resulting in reports from the U.S. Food and Drug Administration (FDA, 2105) and the European Center for Disease Control (ECDC, 2015).

11.2.1 Hazard Analysis

The sources of the outbreaks were water reservoirs of Sorin 3T-heater-coolers in operating rooms (Sax et al., 2015). As the patients did not come in direct contact with water from the heater-cooler, it was proposed and later shown that the route of infection was aerosolization (Sommerstein et al., 2016).

11.2.2 Critical Control Points

Critical control points would include (a) the presence of *M. chimaera* in the Sorin 3T-heater-coolers that were delivered to the hospitals, (b) the filing of the reservoirs with water containing *M. chimaera*, (c) the resistance of *M. chimaera* to the manufacturer's protocol for cleaning and disinfection, (d) the adherence and formation of *M. chimaera* biofilms on the walls of the reservoirs and pipes, pumps, and tubes in the Sorin 3T heater-coolers, (e) the regrowth of *M. chimaera* following disinfection as a result of "dead ends" and biofilms, and (f) the transfer of *M. chimaera* cells from water reservoirs and tubing to the air.

11.2.3 Critical Limits

Critical limits could be established following the identification of critical control points and include (a) the number of *M. chimaera* cells (or other OPPPs) in reservoir water of Sorin 3T heater-coolers (Falkinham, 2020), (b) the number of *M. chimaera* (or other OPPPs) in the water used to fill the heater-cooler reservoirs at hospitals, (c) the number of *M. chimaera* (or other OPPP cells) after following the manufacturer's cleaning and disinfection protocol; possibly sterility or some number per milliliter of water and per square centimeter of biofilm, and (d) the species and number of microorganisms in aerosols collected during operation of the heater-coolers. In this example, the limits for all such samples would be zero, as the patients were at risk and *M. chimaera* is an opportunistic pathogen.

11.2.4 Monitoring Procedures

Monitoring procedures would follow the establishment of critical limits, including (a) chemical and microbiological analysis of the source water used to fill the Sorin 3T heater-coolers, (b) chemical and microbiological analysis of water and biofilm samples from the heater-coolers, (c) collection of aerosol samples (including instrumentation), and (d) documentation of the pathway of microorganisms from the heater-cooler's water reservoir to aerosols, (e) range of airflow velocities, and (f) identification of methods for the detection of aerosol releases (i.e., leaks) during operation of the heater-coolers.

11.2.5 Corrective Actions

Corrective actions would include (a) prior sterilization of Sorin 3T heater-coolers before use, (b) removal of heater-coolers from operating rooms or their isolation within the operating rooms, (c) modification of heater-coolers to prevent the escape of aerosols, (d) HEPA-filtration of exhausted heater-cooler aerosols, (e) filling of heater-cooler reservoirs with sterile water of a defined chemical and microbial composition, (f) increasing disinfectant concentrations or duration of exposure to ensure killing of *M. chimaera*, (g) experiments to measure the efficacy of other disinfectants [e.g., chlorine, chloramine, ozone, ultraviolet irradiation, and high temperature (70 °C)], and (h) identifying the means to kill cells of *M. chimaera* in disinfectant-resistant biofilms.

11.2.6 Record Keeping

Record keeping is straightforward as the monitoring procedures and corrective actions lend themselves to the documentation in laboratory notebooks or online electronic journals; one for each instrument.

11.2.7 Verification

Verification would involve monitoring the frequency of deviations of the measured values over a period of time sufficient to establish averages and standard deviations. Rigorous verification of procedures and critical limits would likely involve inoculation of a heater-cooler with *M. chimaera* to verify the critical control points, the limits, and the procedures.

11.3 Remediation Measures for Water Systems, Buildings, and Homes

As any building's premise plumbing will be colonized by OPPPs and it may prove to be impossible to prevent OPPP colonization of a new building, remediation (the correction of something bad) is necessary. Specifically, the objective is to reduce or eradicate OPPPs in a drinking water system or premise plumbing. There are four levels where management and remediation measures can be imposed: (a) a water provider (utility), (b) a residential building manager, (c) in a hospital, and (d) in a single-family home, apartment, or condominium. The following suggested measures are based on the assumption that drinking water distribution systems and premise plumbing have OPPPs and that OPPP infection is a consequence of exposure to water directly (swallowing) or aerosols. As OPPPs are rather new, caution should be observed in adopting any remediation measures in the following sections, as they are not based on extensive experience of side-by-side trials, but guided by the characteristics of the microorganisms and factors influencing their presence or absence in drinking water. Further, informing premise owners may be problematic. Although the water, residential building, and hospital industries have existing information transfer networks that would lead to wide dissemination for remediation measures, no such network exists for building managers and homeowners. One solution for dissemination of information would be to work through foundations (e.g., NTMir and Cystic Fibrosis Foundations) whose patients are unusually susceptible to infection and disease by OPPPs.

11.3.1 Measures by Water System Operators

In an excellent review "Managing *Legionella pneumophila* in Water Systems" (LeChevallier, 2020) identifies (a) disinfection, (b) hydraulic management (e.g., flushing and cleaning), (c) nutrient limitation, and (d) temperature control (e.g., tank stratification) as control points for managing *L. pneumophila*. Disinfectant residuals should be maintained as OPPPs can grow in the absence of disinfectants. The nutrient reduction would be expected to reduce opportunist plumbing pathogen numbers, as they require low levels of AOC; for example, *M. avium* requires at least 50 μg AOC/L (Norton et al., 2005). Finally, water distribution system pipes could be scoured to reduce biofilms and thereby reduce one source of pathogens; those released from the biofilms.

To Dr. LeChevallier's list, I would add inspection of the distribution system for dead ends where OPPPs persist and grow to re-inoculate the system, and the reduction of turbidity. Turbidity reduction is a standard step of water treatment, resulting in the deposition of solid matter. In addition to reducing particulates in incoming water, a 2-fold drop in *Mycobacterium* spp. in water was observed (Falkinham et al., 2001). That reduction in waterborne *Mycobacterium* spp. numbers reflects the preferential attachment to surfaces, here the particulates. As the OPPPs prefer surface attachment to suspension in water, turbidity reduction should reduce the numbers of all OPPPs.

11.3.2 Measures by Building Owners or Managers

The management measures that could be considered by building owners or managers include (a) identification and removal of all dead ends in pipes, (b) ensuring regular flushing of pipes in areas of low water use to reduce water age and stagnation, (c) installation of whole building filtration, (d) installation of point-of-use filters on taps and faucets, (e) raising water heater temperatures to 130 °F (55 °C), and (e) installation of whole building disinfection (e.g., chlorination or ultraviolet irradiation). These measures would be put in place to prevent or reduce OPPP numbers, not necessarily to treat the water system following an OPPP infection outbreak.

For the remediation where measures are required to reduce OPPP numbers and associated infections, Table 11.1 lists published reports of remediation measures and their efficacy in reducing or eradicating OPPPs. Of the nine publications cited measured the effect of the measure on numbers of OPPPs—two reported results on infection; namely reduction of ICU stay and blood culture infection (Vianelli et al., 2006). All the measures, save a temperature shock at 60 °C (Ji et al., 2018), led to reductions, but not eradication, of OPPP numbers. Point-of-use filtration was uniformly effective for *L. pneumophila* (3-log reduction), with lower reductions for *P. aeruginosa* (2 logs) and *Mycobacterium* spp. (1 log). However, the log-reduction values were limited by the low numbers of *P. aeruginosa* or *Mycobacterium* spp. in the unfiltered water suggesting that higher numbers could have revealed increased efficacy of removal.

Hospitals and residential buildings offer more opportunities for infection by waterborne pathogens (e.g., showers via aerosolization) than office buildings. Prevention and remediation measures could include raising water heater temperature, as there is a lower number of *M. avium* in homes with high hot water heater temperatures (Falkinham, 2011). Further, raising water heater temperatures in 10 homes of *M. avium*-infected patients in Wynnewood, PA showed that by 8–12 weeks *M. avium* disappeared (Lande et al., 2019). Often hospital, healthcare, and multistory residential buildings (e.g., apartments and condominiums) have recirculating hot water systems. As OPPPs can grow and they traverse the pipes, OPPP numbers can be quite high, as high as 100,000 *Mycobacterium* spp. in New York City high-rise buildings (Tichenor et al., 2012). To avoid that problem, building managers could choose to install single unit hot water systems (perhaps instant heaters) to reduce the volume of water heated and the residence time of hot water as it moves through the building.

Table 11.1 Hazard analysis and critical control points analysis

HACCP Step	Action to Reduce OPPPs
Hazard Analysis	OPPPs in premise plumbing OPPPs in instruments OPPPs aerosol generators OPPPs in patients
Critical Control Points	Possible aerosol generators Water heater temperature Disinfectant addition
Critical Limits	Acceptable OPPP limits (water, air) Acceptable OPPP limits (biofilms) Frequency of sampling Concentration of disinfectant
Corrective Action	Thorough cleaning and disinfection Disinfectant addition Remove aerators (aerosol generators)
Record Keeping	Documentation
Verification	Inoculation and testing

Because tap faucet aerators generate aerosols, as do humidifiers and hot tubs (spas) and decorative fountains, they should be removed from buildings. Simply, they serve as sources of aerosolized OPPPs (Hamilton and Falkinham, 2018). Showerheads should be cleaned and disinfected monthly as they are sources of OPPP infection (Falkinham, et al., 2008; Feazel et al., 2009). Here, I make a distinction between cleaning and disinfection. Disinfection means the killing of OPPPs by an agent such as chlorine (e.g., Clorox®). In contrast, cleaning means the removal of surface material and biofilm (e.g., detergent), which may not be free from OPPPs. Further, at-risk individuals could consider replacing a showerhead with one with large holes.

Small holes (< 1 mm) generate mists that remain in the air for a long time and are rich in OPPPs that can enter the farthest reaches of the lung, the alveoli, where OPPPs thrive. Showerheads with large holes (2 mm) generate streams and large drops that quickly fall out of the air.

11.3.3 Management and Remediation in Hospitals and Healthcare Facilities

Hospitals and long-term care facilities present special problems as patients or residents are unusually susceptible to infection. In addition to employing all the measures for suggested residential buildings, the following measures could be considered: (a) removal of sinks and even bathrooms in patient care areas to create a 'water-free' environment (Hopman et al., 2017), (b) reduce aerosols through filter entrapment (e.g., micro-trapping or paraffin-coated), (c) installation of a whole-building disinfection system, and (d) microbial filtration of water in specific locations (e.g., laboratory, equipment preparation, emergency, and operating rooms), where patient trauma will provide open routes of infection.

Filtration to remove OPPPs from water requires filters whose pore size is at least 0.2 μm. Such filters are commercially available and have been shown to prevent exposure to *Legionella* (Sheffer et al., 2005) and reduce the frequency of bacteremia caused by *P. aeruginosa* in a hematology unit (Vianelli et al., 2006).

11.3.4 Management and Remediation by Homeowners

In addition to the measures listed above, the management or remedial action for homeowners is to use a well-water source if possible, as wells have lower numbers of opportunistic pathogens (Falkinham, 2011) and OPPP numbers increase during travel in utility distribution systems from a treatment plant (Falkinham et al., 2001). Further, homeowners should regularly flush water throughout the residence, especially if bedrooms and bathrooms are no longer used regularly. Stagnant water is an ideal habitat for OPPPs and a big concern as office buildings closed during the coronavirus pandemic are reopened. My lab discovered another source of *Mycobacterium abscessus* infection, the chilled water from a refrigerator. Refrigerators can be purchased with an accessible tap for chilled water and ice on the front. However, a woman who regularly drank six 8 oz glasses of water from the refrigerator tap was infected with a strain of *M. abscessus* that had the same fingerprint as did isolates from the refrigerator's water and ice system. Household drinking water was piped to the refrigerator into a 5-gal tank for storage. The tank was outside the chilled compartment and was heated by the exhausted air from the refrigerator cabinet. The refrigerated water had 2 million *M. abscessus* cells per mL!

11.3.5 Speculative Remediation Measures

A number of other management and remediation measures are possible. First, as many of the OPPPs can grow under microaerobic conditions and even prefer microaerobic conditions, hyper-oxygenation might be one way to reduce OPPP numbers. Second, as OPPPs prefer surface biofilms in premise plumbing, the use of anti-adherence or anti-microbial surfaces could reduce OPPP numbers. Several have been developed with demonstrated efficacy in reducing adherent OPPP numbers, namely cuprous oxide (Behzadinasab et al., 2020), shark skin (Kim et al., 2014), and colloidal crystals (Mon et al., 2017). Third, as studies of the microbiome of drinking water have appeared, it has been suggested that investigation of the use of plumbing probiotics be initiated (Wang et al., 2013). The basis for this proposed management and remediation approach is that microorganisms in drinking water may influence the behavior of OPPPs. For example, reduction of amoeba numbers might result in reductions in *L. pneumophila*, due to the possibility that OPPP may require amoeba for growth and survival. Evidence that *Methylobacterium* spp. inhibit adherence and biofilm formation by *M. avium* (Munoz-Egea et al., 2017), suggests that a pipe coating of a *Methylobacterium* spp. cell fraction could prevent the colonization of premise plumbing by *M. avium*.

References

Behzadinasab S, Chin A, Hosseini M, Poon L, Ducker WA. 2020. A surface coating that rapidly inactivates SARS-CoV-2. *ACS Appl Mater Interfaces.* 12: 34723–34727. doi.org/10.1021/acsami.0c11425.

ECDC. 2015. Invasive cardiovascular infection by *Mycobacterium chimaera* potentially associated with heater-cooler units used during cardiac surgery. https://www.ecdc.europa.eu/en/publications-data/invasive-cardiovascular-infection-mycobacterium-chimaera-potentially-associated.

Falkinham JO III. 2011. Nontuberculous mycobacteria from household plumbing of patients with nontuberculous mycobacteria disease. *Emerg. Infect. Dis.* 17: 419–424.

Falkinham JO III. 2020. Disinfection and cleaning of heater-cooler units: Suspension- and biofilm-killing. *J Hosp Infection.* 105: 552–557.

Falkinham JO III, Norton CD, LeChevallier MW. 2001. Factors influencing numbers of *Mycobacterium avium, Mycobacterium intracellulare,* and other mycobacteria in drinking water distribution systems. *Appl Environ Microbiol.* 67: 1225–1231.

Falkinham JO III, Iseman MD, de Haas P, van Soolingen D. 2008. *Mycobacterium avium* in a shower linked to pulmonary disease. *J Water Health.* 6: 209–213.

FDA. 2015. Nontuberculous *Mycobacterium* Infections Associated with Heater-Cooler Devices: FDA Safety Communication. http://www.fda.gov/MedicalDevices/Safety/AlertsandNotices/ucm466963.htm

Feazel LM, Baumgartner LK, Peterson KL, Frank DN, Harris JK, Pace NR. 2009. Opportunistic pathogens enriched in showerhead biofilms. *Proc Natl Acad Sci USA.* 16393–16399. doi: https://doi.org/10.1073/pnas.0908446106.

Hamilton L, Falkinham JO III. 2018. Aerosolization of *Mycobacterium avium* and *Mycobacterium abscessus* from a household ultrasonic humidifier. *J Med Microbiol.* 67: 1491–1497. doi: 10.1099/jmm.0.000822.

Hopman J, Tostmann, A., Wertheim H, Bos M, Kolwijck E, Akkermans R, Sturm P, Voss A, Vickkers P, van der Hoeven H. 2017. Reduced rate of intensive care unit acquired Gram-negative bacilli after removal of sinks and introduction of 'water-free' patient care. *Antimicrob Resist Infect Control.* 6: 59. doi: 10.1186/s13756-017-0213-0. http://ecdc.europa.eu/en/publications/Publications/mycobacterium-chimaera-infection-associated-with-heater-cooler-units-rapid-risk-assessment-30-April-2015.pdf.

Ji P, Rhoads WJ, Edwards MA, Pruden A. 2018. Effect of heat shock on hot water plumbing microbiota and *Legionella pneumophila* control. *Microbiome*. 6: 30. doi: 10.1186/s40168-018-0406-7.

Kim E, Kinney WH, Ovrutsky AR, Vo D, Bai, X, Honda JR, Marx G, Peck E, Lindberg L, Falkinham JO III, May RM, Chan ED. 2014. A surface with a biomimetic micropattern reduces colonization of *Mycobacterium abscessus*. *FEMS Microbiology Letters*. 359: 1–6.

Lande L, Alexander DC, Wallace RJ Jr, Kwait R, Iakhiaeva E, Williams M, Cameron ADS, Olshefsky S, Devon R, Vasireddy R, Peterson DD, Falkinham JO III. 2019. *Mycobacterium avium* in community and household water, suburban Philadelphia, Pennsylvania, USA, 2010–2012. *Emerg Infect Dis*. 25: 473–481.

Lechevallier MW. 2020. Managing *Legionella pneumophila* in water systems. *J Am Water Works Assoc*. 112: 10–23.

Mon H, Chang Y-R, Ritter AL, Falkinham JO III, Ducker W. 2017. Effects of colloidal crystals, antibiotics, and surface-bound antimicrobials on *Pseudomonas aeruginosa* surface density. *ACS Biomater Sci Eng*. 4: 257–265.

Muńoz-Egea MC, Ji P, Pruden A, Falkinham JO III. 2017. Inhibition of adherence of *Mycobacterium avium* to plumbing surface biofilms of *Methylobacterium* spp. *Pathogens*. 6: 42.

Norton CD, LeChevallier MW, Falkinham JO III. 2004. Survival of *Mycobacterium avium* in a model distribution system. *Water Res*. 38: 1457–1466.

Rhodes WJ, Pruden A, Edwards MA. 2014. Anticipating challenges with in-building disinfection for control of opportunistic pathogens. *W Environ Res*. 86: 540–549.

Saidan MN, Abdalla AI, Al Alami N, Al-Naimat H. 2019. Multiple disinfection processes of *Legionella pneumophila* positive in hotels' water distribution systems in Jordan. *Desal*. 163: 7–16. doi: 10.5004/dwt.2019.24411.

Sax H, Bloemberg G, Hasse B, Sommerstein R, Kohler P, Achermann Y, Rössle M, Falk V, Kuster SP, Böttger EC, Weber R. 2015. Prolonged outbreak of *Mycobacterium chimaera* infection after open-chest heart surgery. *Clin Infect Dis*. **61:** 67–75. doi: 10.1093/cid/civ198.

Sheffer PJ, Stout JE, Wagener MM, Muder RR. 2005. Efficacy of new point-of-use water filter for preventing exposure to *Legionella* and waterborne bacteria. *Am J Infect Control*. 33: S20–S25. doi: 10.1016/ajic.2005.03.012.

Sommerstein R, Rüegg C, Kohler P, Bloemberg G, Kuster BP, Sax H. 2016. Transmission of *Mycobacterium chimaera* from heater-cooler units during cardiac surgery despite an ultraclean air ventilation system. *Emerg Infect Dis.* 22: 1008–1013.

Tichenor WS, Thurlow J, McNulty S, Brown-Elliott BA, Wallace RJ Jr, Falkinham JO III. 2012. Nontuberculous mycobacteria in household plumbing as possible cause of chronic rhinosinusitis. *Emerg Infect Dis.* 18: 1612–1617.

Totaro M, Mariotti T, Bisordi C, De Vita E, Valentini P, Costa AL, Casini B, Privitera G, Baggiani A. 2020. Evaluation of *Legionella pneumophila* decrease in hot water network of four hospital buildings after installation of electron time flow taps. *Water.* 12: 210. doi: 10.3389/w12010210.

Vianelli N, Giannini MB, Quarti C, Sabattini MAB, Flacchini M, de Vivo A, Graldi P, Galli S, Nanetti A, Baccarani M, Ricci P. 2006. Resolution of a *Pseudomonas aeruginosa* outbreak in a hematology unit with the use of disposable sterile water filters. *Hematologica.* 91: 983–985.

Wang H, Edwards MA, Falkinham JO III, Pruden A. 2013. Probiotic approach to pathogen control in premise plumbing systems: A review. *Environ Sci Technol.* 47: 10117–10128.

Whiley H, Bentham R, Brown MH. 2017. *Legionella* persistence in manufactured water systems: Pasteurization potentially selecting for thermal tolerance. *Front Microbiol.* 8: 1330; doi: 10.3389/fmicb.2017.01330.

Williams MM, Chen T-H, Keane T, Toney N, Toney S, Armbruster CR, Butler WR, Arduino MJ. 2011. Point-of-use membrane filtration and hyperchlorination to prevent patient exposure to rapidly growing mycobacteria in the potable water supply of a skilled nursing facility. *Infect Control Hosp Epidemiol.* 32: 837–844. doi: 10.1086;661282.

Zhou ZY, Hu BJ, Qin L, Lin YE, Watanabe H, Zhou Q, Gao XD. 2013. Removal of waterborne pathogens for liver transplant unit water taps in prevention of healthcare-associated infections: A proposal for a cost-effective, proactive infection control strategy. *Clin Microbiol Infect.* 20: 310–314. doi: 10.1111/1469-0691.12299.

Case Studies

Case Study 1. "Patients Teach You to Sample Refrigerator Taps"

In 2008, I received a call from an engineer who asked if my lab could sample his mother's house for mycobacteria as she was infected with *Mycobacterium abscessus*. As we were in the midst of a study of mycobacteria in patients' homes supported by a grant from the Nontuberculous Mycobacteria Information and Research Foundation (www.ntminfo.org), I readily agreed and sent a collection kit. After receipt of the samples and their processing, my technician showed me a plate with more than 10,000 colonies of *M. abscessus*. As we spread 0.1 mL samples, the original sample had 100,000 *M. abscessus* cells per milliliter. That was the highest number we had seen from a household sample; more than 10,000-times higher. We looked at the list of samples and discovered that the sample was water from the refrigerator tap on the front. We had never thought to sample water from a refrigerator tap! I called the engineer's son and asked about the sample. He confirmed that it came from his mother's refrigerator (chilled) and further told me that his mother drank six 8-ounce glasses of water daily. As there are 1,420 mL per ounce and the mother drank a total of 48 ounces daily, I calculated that she was drinking 68,000 mL. That, in turn, meant she was drinking over 6 billion *M. abscessus* cells per day. The engineer son immediately stopped his mother from drinking water from the refrigerator tap. He even removed and sent the entire system for the water to my lab. It was loaded with *M. abscessus*. Now you know why I listen carefully to patients and tell everyone not to drink refrigerator tap water.

Case Study 2. "Patients Teach You to Sample Showerheads"

In 2007, I received a call from Dr. Mike Iseman at National Jewish Hospital. He was treating a woman with *Mycobacterium avium* lung disease who thought she had been infected from her shower. That was a novel source that we had not sampled previously, so I sent a collection kit to the woman. She sent samples of her cold and hot water as well as sediment from the showerhead. After cultivation and sufficient time for the slow-growing mycobacteria to grow and form colonies (14–21 days), we identified high numbers of *M. avium* in the shower and hot water samples. Through the efforts of Dr. Iseman, we received a culture of the patient's lung isolate and were able to show that its DNA fingerprint was identical to several of the isolates from her showerhead and hot water. As in case study number 1, this experience taught me to listen to patients as they can provide clues to identifying the source of their infecting pathogen.

Case Study 3. "Read Everything in Published Scientific Papers, or Pink Slime in the Shower Is Not All Bad"

In 2008, I had been in Denver and a colleague, Dr. Mary Ann DeGroote asked if I would travel up to Boulder and speak to Dr. Norman Pace's research group about mycobacteria in showerheads. I was delighted and spent an afternoon at Dr. Pace's lab at the University of Colorado. I was flattered that they listened to my showerhead story and within a year, I received a draft of a manuscript from Norm's lab describing their success in demonstrating the widespread presence of mycobacteria in showerheads across the United States. Dr. Pace's group could do that quickly as they isolated and analyzed DNA and did not have to wait for the slow growth of mycobacteria. When I saw the published paper, I was struck by the fact that mycobacteria were the most frequently detected bacteria in showerheads. The second most common was methylobacteria, salmon pink-pigmented waterborne bacteria. I was so interested in their paper that I downloaded and went through the supplementary data; sample-by-sample. What I discovered was that although both bacteria were frequent, they very rarely were found in the same sample; if methylobacteria were present, mycobacteria were absent and vice-versa. At that time, we were engaged in a study of plumbing in homes of *M. avium*-infected individuals living in a suburb of Philadelphia. We confirmed what Dr. Pace's research group had studied. Those observations form the basis for my advice that if you have a salmon-pink slime on shower tiles, grout, glass, or curtains, it is likely that mycobacteria are absent. That fact is important to individuals who are at risk of mycobacterial lung disease. Later work in my lab showed that methylobacteria adhering to surfaces prevents the adherence of mycobacteria. By preventing adherence, the slowly growing mycobacteria cannot persist in plumbing, they get washed out. Because of that experience, I spend more time reading the supplementary data in published papers.

Case Study 4. "*Legionella pneumophila* Where You Shop"

In October 1996, those living in Blacksburg, where Virginia Tech is located, learned about a local outbreak of Legionnaire's disease. The New River Valley Health District passed the word along of the outbreak to the Virginia Department of Health. Further, the increases in pneumonias due to unidentified organisms occurred at more than one hospital and *L. pneumophila* antibodies were identified in a subset of patients it was felt that there was an outbreak. Interviews with patients led to the discovery that 14 of 15 case-patients had visited a large home-improvement store. Only 12 of 45 control individuals had visited that store. Further, case-patients had stayed an average of 79 minutes, whereas controls averaged 29 minutes. Case patients had also spent time in an area with demonstration spas and samples of a filter from one of the spas yielded *L. pneumophila* of the same monoclonal antibody type and DNA fingerprint generated by arbitrarily primed PCR. Although I was aware that the human environment can harbor *L. pneumophila*, this outbreak focused my attention on hot tubs as a possible source of nontuberculous mycobacteria and other opportunistic premise plumbing pathogens.

Case Study 5. "Getting Reacquainted an Old Friend, *Cupriavidus*"

In July 2019, a professional friend, Michael Weintraub, a consulting epidemiologist sent me a note asking if I had any experience with *Cupriavidus*. I had but in a completely different context than the report, Mike brought to my attention. It was a report of an outbreak of bloodstream infections caused by *Cupriavidus pauculus* in a haemato-oncology unit at Queen Elizabeth University Hospital in Glasgow, Scotland. Although I did know about *Cupriavidus*, but not as responsible for human infection. At Glasgow Hospital, the infecting *C. pauculus* strain was found throughout the hospital's water system. I knew about it as it had been described as a predator bacterium by Lester Casida at Pennsylvania State University. *Cupriavidus* spp. (i.e., copper-loving) were named because they kill other bacteria by extracting copper. They are waterborne environmental bacteria and now I place them in the category of opportunistic premise plumbing pathogens (OPPPs) based on the outbreak in Glasgow and other reports of infection. I suspect their residence in hospital plumbing is due, in part, to their acquisition of copper.

Case Study 6. "Why Are OPPPs Intracellular Pathogens?"

L. pneumophila, P. aeruginosa, M. avium complex, Acinetobacter baumannii, Stenotrophomonas maltophilia, and *Methylobacterium* spp. belong to a group called, amoeba-resisting microorganisms (ARMS). These are bacteria that can survive the predation of amoeba. They even grow in amoeba. Thus, the amoeba serves as a site for OPPP amplification. It is not surprising that OPPPs are ARMs. The OPPPs are waterborne opportunists and widely distributed in natural waters and in human-engineering water systems, including premise plumbing. Normally, amoeba grazes on bacteria in biofilms. Thus, OPPPs are under strong selective pressure to evade being killed upon phagocytosis by amoeba. Their ability to survive and grow in amoeba also leads to their ability to persist and grow in human macrophages. Further, we discovered another link between amoeba and *M. avium*. Earlier, a student in my lab showed that *M. avium* could survive and grow in the protozoan, *Tetrahymena pyriformis*. I went to the library and spent a day in the protozoa section (i.e., *Journal of Protozology*) and found a report that *Tetrahymena* spp. failed to grow from low inoculum densities unless the medium was supplemented with fatty acids, pyrimidines, or other compounds. Now, if the mycobacteria are anything, they are rich sources of fatty acids and I hypothesized that mycobacterial cells could substitute for fatty acids and trigger growth of *T. pyriformis* from low cell density. We then measured the growth of *T. pyriformis* cells that either carried or did not carry intracellular *M. avium*. If the hypothesis was correct, only the *M. avium*-infected *T. pyriformis* cells would grow from low cell density. That turned out to be the case and now we have another reason for the association of *M. avium* and *T. pyriformis*, growth stimulation. Perhaps a similar relationship exists between amoeba and the other OPPPs.

Case Study 7. "Energy Savings at Odds with Public Health"

In our initial study of the presence of nontuberculous mycobacteria (NTM) in patient houses—sponsored by the NTMir Foundation (www.ntminfo.org)—the patients filled out a questionnaire about their household plumbing. One of the questions asked was to provide us with the set temperature of the household water heater and whether the water was heated by gas or electricity. I felt that might be a value as others had reported an association between *Legionella* numbers and water heater type and temperature.

After we had accumulated the data on the presence of NTM, we discovered that the frequency of isolation of *Mycobacterium* spp. was higher in homes with water heaters set at 125 °F (50 °C) or below. If the water heater set temperature was 130 °F (55 °C), we seldom found any *Mycobacterium* spp. As a result of that, we recruited 10 *M. avium*-infected patients in a suburb of Philadelphia to raise their water heater temperatures to 130 °F or higher. In all 10 homes, *M. avium* disappeared by 8–12 weeks. We continue to monitor the presence of *M. avium* in those homes as we realize that strong selective pressure is being applied and that selection may lead to the emergence of temperature-resistant *M. avium*.

Now, that data places me firmly against the recommendation of a variety of public government agencies and consumer groups to reduce energy usage by reducing the temperature of water heaters. Further, it raises questions about the wisdom of "Green Buildings". In such buildings, the temperature of heated water is lower. Simply, I see a conflict between our desire to reduce energy consumption and the safety of individuals at risk for *Mycobacterium* spp. pulmonary disease.

Index